GCSE 9-1

HIGHER

MATHEMATICS

EDEXCEL

REVISION AND EXAM PRACTICE

Stephen Doyle

Author Stephen Doyle
Editorial team Haremi Ltd
Series designers emc design ltd
Typesetting York Publishing Solutions Pvt. Ltd.
Illustrations York Publishing Solutions Pvt. Ltd.
App development Hannah Barnett, Phil Crothers and Haremi Ltd

Designed using Adobe InDesign
Published by Scholastic Education, an imprint of Scholastic Ltd, Book End, Range Road, Witney, Oxfordshire, OX29 0YD
Registered office: Westfield Road, Southam, Warwickshire CV47 0RA
www.scholastic.co.uk

Printed by Bell & Bain Ltd, Glasgow
© 2017 Scholastic Ltd
1 2 3 4 5 6 7 8 9   7 8 9 0 1 2 3 4 5 6

British Library Cataloguing-in-Publication Data
A catalogue record for this book is available from the British Library.
ISBN 978-1407-16900-2

Notes from the publisher

Please use this product in conjunction with the official specification and sample assessment materials. Ask your teacher if you are unsure where to find them.

The marks and star ratings have been suggested by our subject experts, but they are to be used as a guide only.

Answer space has been provided, but you may need to use additional paper for your workings.

How to use this book

Inside this book you'll find everything you need to help you succeed in the new GCSE 9–1 Edexcel Higher Mathematics specification. It combines revision and exam practice in one handy solution. Broken down into topics and subtopics, it presents the information in a manageable format. Work through the revision material first or dip into the exam practice as you complete a topic. This book gives you the flexibility to revise your way!

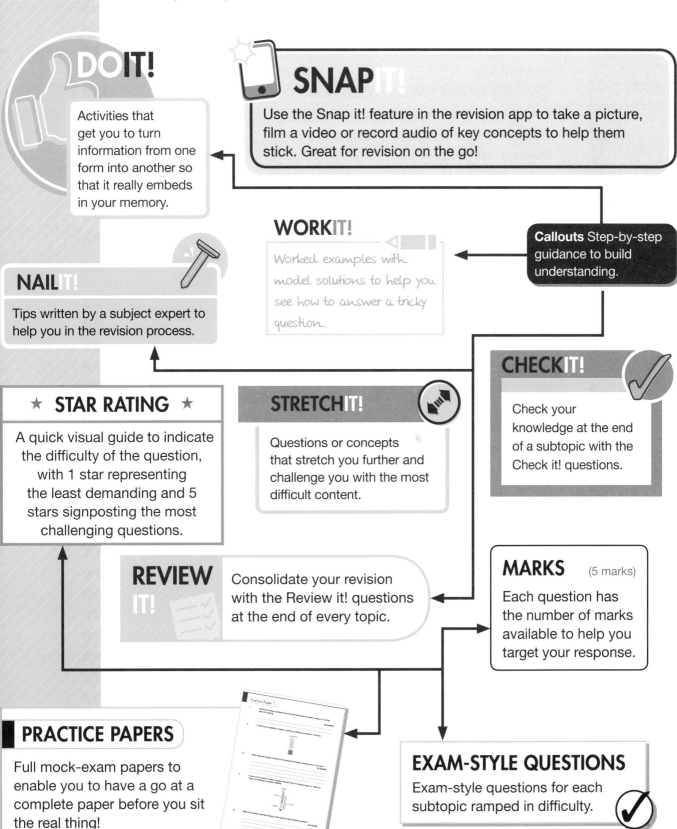

DOIT!

Activities that get you to turn information from one form into another so that it really embeds in your memory.

SNAPIT!

Use the Snap it! feature in the revision app to take a picture, film a video or record audio of key concepts to help them stick. Great for revision on the go!

Callouts Step-by-step guidance to build understanding.

WORKIT!

Worked examples with model solutions to help you see how to answer a tricky question.

NAILIT!

Tips written by a subject expert to help you in the revision process.

CHECKIT!

Check your knowledge at the end of a subtopic with the Check it! questions.

★ STAR RATING ★

A quick visual guide to indicate the difficulty of the question, with 1 star representing the least demanding and 5 stars signposting the most challenging questions.

STRETCHIT!

Questions or concepts that stretch you further and challenge you with the most difficult content.

REVIEW IT!

Consolidate your revision with the Review it! questions at the end of every topic.

MARKS (5 marks)

Each question has the number of marks available to help you target your response.

PRACTICE PAPERS

Full mock-exam papers to enable you to have a go at a complete paper before you sit the real thing!

EXAM-STYLE QUESTIONS

Exam-style questions for each subtopic ramped in difficulty.

Revision contents

Topic 4

GEOMETRY AND MEASURES

Topic 5

PROBABILITY

Topic 6

STATISTICS

Exam Practice contents

Topic 4   GEOMETRY AND MEASURES

Topic 5   PROBABILITY

Topic 6   STATISTICS

HOW TO REVISE!

PLAN YOUR REVISION

Get ahead by planning your revision!

Work out the **time** you have available for revising.

Think about when you work at your best. Are you a morning or an evening person?

Allocate **MORE TIME** for the topics you struggle with.

Revision works best in **SMALL BURSTS**, so keep sessions **SHORT AND SWEET**!

Remember to allow time to **PRACTISE** applying what you have revised.

Use your **revision app** to put together a revision timetable.

LOOK AFTER YOURSELF

Help your brain by looking after your whole body!

Take regular **breaks** from revising – your brain needs time to digest information in order to retain it.

HOTEL

Keep **hydrated** by drinking plenty of water – dehydration stops your brain from working at its full capacity.

Regular **exercise** helps stimulate the brain and will help you relax.

Get plenty of **sleep**, especially the night before an exam.

EAT WELL and limit unhealthy snacks – your brain needs fuel for memory and concentration.

Find methods of **relaxation** that work for you throughout the revision period.

BE PREPARED!

Limit potential stress on the day of an exam by getting everything you need ready the night before.

30

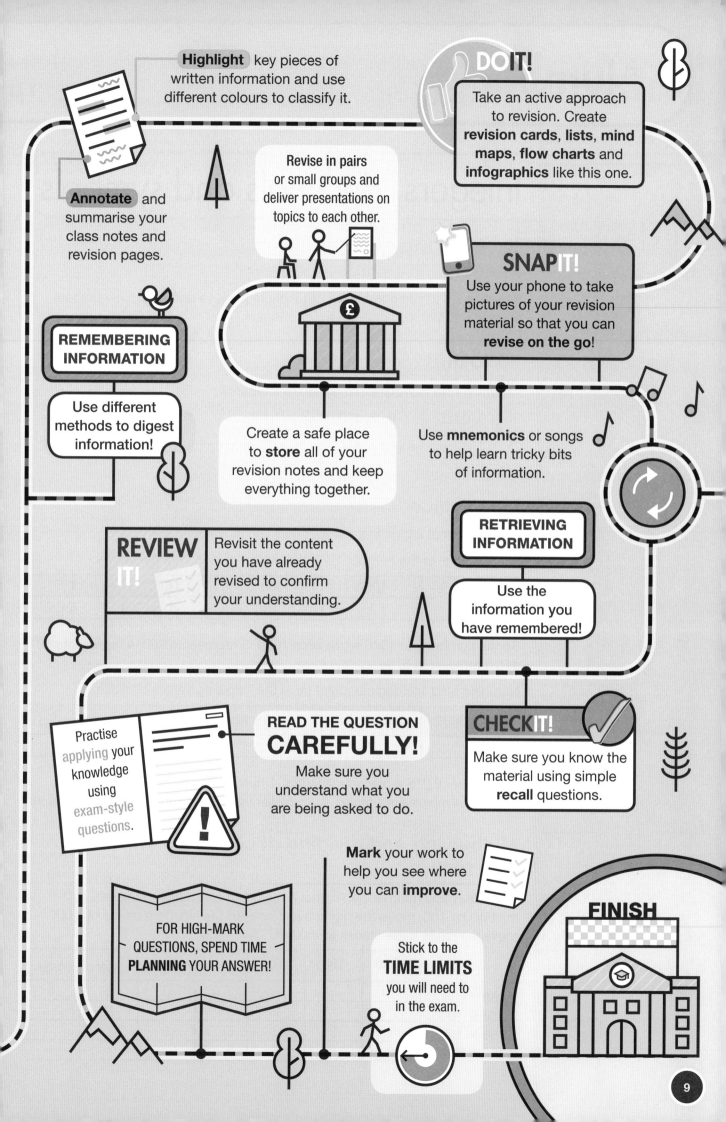

Highlight key pieces of written information and use different colours to classify it.

DO IT!
Take an active approach to revision. Create **revision cards**, **lists**, **mind maps**, **flow charts** and **infographics** like this one.

Annotate and summarise your class notes and revision pages.

Revise in pairs or small groups and deliver presentations on topics to each other.

SNAP IT!
Use your phone to take pictures of your revision material so that you can **revise on the go!**

REMEMBERING INFORMATION

Use different methods to digest information!

Create a safe place to **store** all of your revision notes and keep everything together.

Use **mnemonics** or songs to help learn tricky bits of information.

REVIEW IT!
Revisit the content you have already revised to confirm your understanding.

RETRIEVING INFORMATION

Use the information you have remembered!

Practise applying your knowledge using exam-style questions.

READ THE QUESTION CAREFULLY!
Make sure you understand what you are being asked to do.

CHECK IT!
Make sure you know the material using simple **recall** questions.

Mark your work to help you see where you can **improve**.

FINISH

FOR HIGH-MARK QUESTIONS, SPEND TIME **PLANNING** YOUR ANSWER!

Stick to the **TIME LIMITS** you will need to in the exam.

Number

Integers, decimals and symbols

Integers

Integers are whole numbers. They include positive and negative numbers and zero. 1, 129, 0, −2, −32 are all integers.

WORKIT!

Which **four** numbers in this list are integers?

$\frac{1}{4}$, 0, 0.125, $\frac{1}{3}$, −0.05, 120, π, $\sqrt{4}$, $\sqrt{2}$, 17

The four integers are 0, 120, $\sqrt{4}$, 17.

> $\sqrt{4}$ = 2 (or −2) and is therefore an integer.

Place value

The position of a digit in a number tells you its value.

For example, in the number 1346 the values of digits are as follows:

Place value	1000	100	10	1
Number	1	3	4	6

So the number 1 in 1346 represents 1000, 3 represents 300, 4 represents 40 and 6 represents 6.

Numbers after the decimal point have the following values of digits:

Place value	tenth $\left(\frac{1}{10}\right)$	hundredth $\left(\frac{1}{100}\right)$	thousandth $\left(\frac{1}{1000}\right)$
Number	2	1	5

The number 0.215 represents 2 tenths, 1 hundredth and 5 thousandths. The number 0.4 represents 4 tenths. 0.4 is bigger than 0.215 because it has more tenths.

Multiplying and dividing decimals by 10, 100 and 1000

To multiply any number by 10, move the digits one place to the left. To multiply by 100, move the digits two places to the left. To multiply by 1000, move the digits three places to the left.

	1000	100	10	1	.	$\frac{1}{10}$	$\frac{1}{100}$	$\frac{1}{1000}$
2.358				2	.	3	5	8
2.358 × 10			2	3	.	5	8	
2.358 × 100		2	3	5	.	8		
2.358 × 1000	2	3	5	8	.			

Similarly, to divide by 10, 100 or 1000, move the digits one, two or three places to the right.

	1000	100	10	1	.	$\frac{1}{10}$	$\frac{1}{100}$	$\frac{1}{1000}$
70			7	0				
70 ÷ 10				7				
70 ÷ 100				0	.	7		
70 ÷ 1000				0	.	0	7	
1502	1	5	0	2				
1502 ÷ 10		1	5	0	.	2		
1502 ÷ 100			1	5	.	0	2	
1502 ÷ 1000				1	.	5	0	2

> If necessary, put zeros into the 'empty' places after the decimal point.

WORKIT!

$0.37 \times 35 = 12.95$

Without using a calculator, work out

a 37×35

> $37 = 100 \times 0.37$, so the answer will be 100 times bigger. Move the digits in 12.95 two places to the left.

  $37 \times 35 = 100 \times 0.37 \times 35$

  $= 12.95 \times 100 = 1295$

b 3.7×350

  $3.7 \times 350 = 100 \times 0.37 \times 35$

  $= 12.95 \times 100 = 1295$

> $3.7 = 10 \times 0.37$ and $350 = 10 \times 35$, so the answer will be $10 \times 10 = 100$ times bigger.

c $\dfrac{12.95}{350}$

  $0.37 = \dfrac{12.95}{35}$

> Divide both sides of the original calculation by 35.

  $\dfrac{12.95}{350} = \dfrac{12.95}{35 \times 10} = 0.37 \div 10 = 0.037$

> $350 = 10 \times 35$. Since the 350 is the denominator, the answer will be 10 times smaller.

d $\dfrac{1295}{37}$

  $\dfrac{12.95}{0.37} = 35$

  $\dfrac{1295}{37} = \dfrac{12.95 \times 100}{0.37 \times 100} = 35$

> $1295 = 100 \times 12.95$ and $37 = 100 \times 0.37$, and the two 100s cancel each other out.

Mathematical symbols

Mathematical symbols save time when writing. You need to remember the following symbols.

SNAP IT!   Mathematical symbols

Symbol	Meaning	Example
$=$	is equal to	$x + x = 2x$
$\neq$	is not equal to	$1 + 5 \times 2 \neq 12$
$>$	is greater than	$-3 > -4$
$<$	is less than	$0 < 4$
$\geq$	is greater than or equal to	$x \geq 4$
$\leq$	is less than or equal to	$y \leq 6$

DOIT!

Put each symbol and its meaning on a stickynote and place them around your house to help you remember them.

We don't usually write
the sign unless it is
negative.

Directed numbers

Numbers can be positive or negative or zero. Numbers with a sign
(i.e. positive or negative numbers) are called **directed numbers**.
Numbers therefore have a size and a sign (i.e. + or −). Directed
numbers can go above and below a zero value. You can see this
with temperature, which can be positive, negative or zero.

Thinking of directed numbers in terms of temperature

A good way to deal with directed numbers is to think of a
thermometer. A thermometer has positive values above zero and
negative values below zero.

To work out $-3 + 6$, start at the first number (-3). The next number
($+6$) means that you will go up 6 units. The first 3 units of the 6 will take you to zero
and the next 3 units will take you 3 units above zero to give the answer $+3$ or just 3.

To work out $-1 - 4$, start at -1 and go down 4 units (because the sign in
front of the 4 is negative): $-1 - 4 = -5$.

Calculations involving more than two directed numbers

To work out the value of (i.e. evaluate) $-5 + 4 - 3$:
- find out the result of the first two numbers: $-5 + 4 = -1$
- then use the result with the final number: $-1 - 3 = -4$.

SNAP IT!   Multiplication and division of directed numbers

$+ \times + = +$	$- \times - = +$
$+ \times - = -$	$- \times + = -$
$+ \div + = +$	$- \div - = +$
$+ \div - = -$	$- \div + = -$

Multiplication and division of directed numbers

The rules for the multiplication of directed
numbers are that if the two numbers being
multiplied/divided are both the same sign (i.e.
both positive or both negative), the answer
will be positive. If the two numbers have
opposite signs, the answer will be negative.

CHECK IT!

1 $23 \times 0.087 = 2.001$
Without using a calculator, work out
a 23×8.7
b 2.3×0.87
c $\dfrac{2.001}{0.87}$
d $\dfrac{2001}{23}$

2 $486 \times 29 = 14094$
Without using a calculator, work out
a 4.86×29
b 0.486×2.9
c $14094 \div 48.6$
d $140.94 \div 29$

3 Arrange these numbers in ascending order.

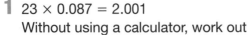

0.12   0.012   0   −0.5   12

4 Place the correct symbol from the following
list in the box. Each symbol can be used
once, more than once or not at all.

$$< \quad > \quad \geq \quad \leq \quad =$$

a $\dfrac{5}{0.5} \, \square \, 10$ **b** $1\dfrac{5}{9} \, \square \, \dfrac{4}{3}$ **c** $-3 \, \square \, -1$

5 Work out
a $-2 + 7$
b $-3 - 5$
c 3×-5
d $-12 \div -2$
e $-1 + 7 - 10$

Addition, subtraction, multiplication and division

Addition

This is the same as writing numbers in place value tables.

To add integers:

1. Write the numbers one above the other, with the units in the same column, i.e. lined up to the right.

2. Starting with the column on the right, add up each column of numbers.
   If the total is more than 10, write the tens digit under the next column to the left.

3. Repeat for each column. If there is a digit carried from the previous column, include this in the addition.

Similarly, to add decimal numbers, line up the decimal points vertically.

WORKIT!

Find the sum of the numbers 152, 45 and 1723.

```
  1 7 2 3
    1 5 2
+     4 5
  1 9 2 0
      1 1
```

3 + 2 + 5 = 10. Write 0 in the first column and carry 1 into the next column.

Make sure that you write the carry digit in the next column to the left. Remember to add it to the total of that column.

WORKIT!

Work out 97.56 + 4.054.

```
   9 7 · 5 6
 + 4 · 0 5 4
 1 0 1 · 6 1 4
       1     1
```

Line up the decimal points.

Subtraction

To subtract integers, line the numbers up as you did for addition. Again, start with the column on the right (the units column). In each column, always subtract the number on the bottom row from the number on the top row.

The units column is a more formal name for the ones column.

WORKIT!

Subtract 129 from 647.

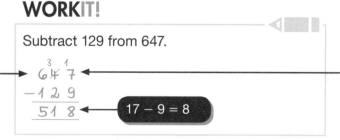

```
    3  1
  6 4 7
- 1 2 9
  5 1 8
```

10 has been added to the 7, so reduce the number in the tens column by 1. Cross out the 4 and write 3.

9 is larger than 7, so 'borrow' from the next column to the left. Write 1 by the 7 to make 17.

17 − 9 = 8

To subtract decimals, line the numbers up on the decimal point. If the number of decimal places in the two numbers is different, fill the empty places with zeros.

NAIL IT!

You may have been taught to write this slightly differently. Use the method you are most comfortable with.

WORKIT!

Work out 68.03 − 5.428.

$$
\begin{array}{r}
6\,\overset{7}{8}\cdot\overset{1}{0}\,\overset{2}{3}\,\overset{1}{0}\\
-\quad 5\cdot 4\ 2\ 8\\
\hline
6\,2\cdot 6\ 0\ 2
\end{array}
$$

Write 0 at the end of 68.03 so both numbers have 3 decimal places.

Subtract each number on the bottom row from each number on the top row. Borrow from the next column to the left when necessary.

WORKIT!

Without using a calculator, multiply 315 by 27.

$$
\begin{array}{r}
3\ 1\ 5\\
\times\ 2\ 7\\
\hline
2\,2_1 0_3 5\\
6\,3_1 0\ 0\\
\hline
8\ 5\ 0\ 5
\end{array}
$$

315 × 7 = 2205

315 × 2 = 630, so 315 × 20 = 6300

2205 + 6300 = 8505

Multiplication

To multiply integers:

1. Write the numbers one above the other, with the units in the same column, i.e. lined up to the right.

2. Multiply the units digit in the second integer by the first integer. Write this under the line.

3. Multiply the tens digit in the second integer by the first integer. As this is the tens digit, you need to multiply the answer by 10. Write this in the next line.

4. Repeat for higher place digits in the second integer.

5. Add the results of steps 2 to 4.

To multiply decimals, for example 6.12 × 7.4:

1. Multiply the numbers without the decimal points: 612 × 74 = 45 288.

2. Count the number of decimal places in each number: 2 and 1.

3. Add these together: 3.

4. Insert the decimal point in the answer to step 1 so that the number has the number of decimal places found in step 3: 45.288.

WORKIT!

Without using a calculator, work out 7.8 × 4.1.

$$
\begin{array}{r}
7\ .\ 8\\
\times\ 4\ .\ 1\\
\hline
7\ \ 8\\
3\ 1_3\ 2\ \ 0\\
\hline
3\ 1\ .9\ \ 8
\end{array}
$$

7.8 and 4.1 each have 1 decimal place, so total number of decimal places = 2. Insert the decimal point in the answer.

Multiply the numbers as if they were 78 and 41. Add up the two results.

NAIL IT!

Check your answer by estimating: 8 × 4 = 32, so the decimal point is in the right place.

Division

WORKIT!

Without using a calculator, work out 459 ÷ 17.

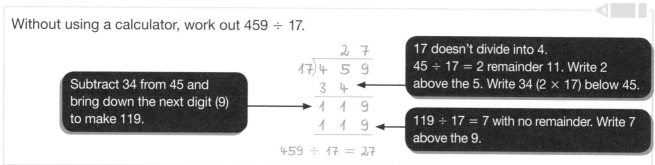

$$459 \div 17 = 27$$

Subtract 34 from 45 and bring down the next digit (9) to make 119.

17 doesn't divide into 4. 45 ÷ 17 = 2 remainder 11. Write 2 above the 5. Write 34 (2 × 17) below 45.

119 ÷ 17 = 7 with no remainder. Write 7 above the 9.

To divide by a decimal, multiply both numbers by 10 (or 100, 1000,…) until you are dividing by a whole number.

WORKIT!

Without using a calculator, work out 55.3 ÷ 3.5.

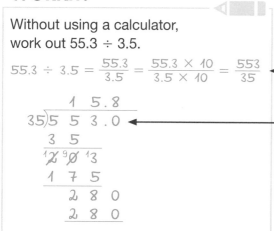

$$55.3 \div 3.5 = \frac{55.3}{3.5} = \frac{55.3 \times 10}{3.5 \times 10} = \frac{553}{35}$$

Multiply both numbers by 10, so you are dividing by 35.

Insert a decimal point and zeros after the number if necessary.

DOIT!

Make a revision card for each type of calculation.

CHECKIT!

1 Without using a calculator, work out

  a 1083 + 478

  b 2445 + 89 + 513

  c 66.55 + 3.38

  d 7.08 + 4.5 + 12.343

2 Without using a calculator, work out

  a 4556 − 1737   **c** 12.935 − 4.75

  b 674 − 387   **d** 5.77 − 0.369

3 Without using a calculator, work out

  a 634 × 47   **c** 8.32 × 4.9

  b 7.7 × 3.8

4 Without using a calculator, work out

  a 1058 ÷ 23   **c** 88.5 ÷ 2.5

  b 617.4 ÷ 1.8

Using fractions

Fractions are one number divided by another number, written as one over the other. Here are some terms that may be used in fraction questions. You need to remember these.

SNAP IT! Types of fraction

- **Proper fractions** are fractions where the top number (the numerator) is smaller than the bottom number (the denominator), for example $\frac{2}{3}$, $\frac{1}{2}$ and $\frac{15}{16}$.

- **Improper fractions** are fractions where the top number (the numerator) is bigger than the bottom number (the denominator), for example $\frac{4}{3}$, $\frac{16}{5}$ and $\frac{36}{4}$.

- **Mixed numbers** are numbers consisting of an integer and a fractional part, for example $5\frac{4}{5}$.

Cancelling fractions

Fractions should be cancelled, if possible, by dividing the numerator and denominator by the same number that divides in exactly. Keep doing this until the fraction can't be cancelled any further.

NAIL IT!

'Simplify' means to cancel the fraction until you can't cancel it any more.

Try to pick the largest number you can as the cancelling will be quicker. For example, 6 divides exactly into both 12 and 18, taking you quickly to the answer.

WORKIT!

Simplify the fraction $\frac{12}{18}$.

$$\frac{12}{18} = \frac{6}{9} = \frac{2}{3}$$

Find a number which divides exactly into the numerator and denominator. Keep going until you cannot divide by any more numbers.

Converting an improper fraction to a mixed number

To convert an improper fraction such as $\frac{5}{3}$ to a mixed number:

1 Divide the denominator (bottom number) into the numerator (top number) and write the integer down: 3 divides into 5 once with a remainder of 2, so write down 1.

2 Write the remainder over the denominator and write this next to the integer: $\frac{5}{3} = 1\frac{2}{3}$.

Converting a mixed number to an improper fraction

You often have to do this when you are multiplying or dividing fractions. To convert a mixed number such as $3\frac{2}{5}$ to an improper fraction:

1 Multiply the whole number by the denominator of the fraction: $3 \times 5 = 15$.

2 Add the numerator: $15 + 2 = 17$.

3 Put the result over the denominator: $\frac{17}{5}$.

Calculations involving fractions

Calculations involving fractions are easy if you obey some rules and are careful.

Adding fractions

You can add two fractions if they have **the same** denominator (i.e. the same number on the bottom).

- If the bottom numbers are the same, add the top numbers and then cancel if possible. For example, $\frac{5}{16} + \frac{7}{16} = \frac{12}{16} = \frac{3}{4}$.

- If the bottom numbers are **not the same**, find the common denominator (the smallest number both denominators divide exactly into).

> Both 5 and 3 divide exactly into 15. Change the first fraction's denominator to 15 by multiplying the top and bottom by 3. Change the second fraction's denominator to 15 by multiplying the top and bottom by 5.

WORKIT!

Work out $\frac{3}{5} + \frac{1}{3}$.

$$\frac{3}{5} + \frac{1}{3} = \frac{9}{15} + \frac{5}{15} = \frac{14}{15}$$

> Once the denominators are the same, add the top numbers.

Adding two mixed numbers

To add two mixed numbers:

1. Add the whole numbers.

2. Add the fractions.

3. If the fractions add to an improper fraction, then convert to a mixed number.

4. Add your answer from step 1 to your answer from step 3.

WORKIT!

Work out $3\frac{3}{4} + 4\frac{3}{5}$.

$$3\frac{3}{4} + 4\frac{3}{5} = 7 + \frac{3}{4} + \frac{3}{5}$$
$$= 7 + \frac{15}{20} + \frac{12}{20}$$
$$= 7\frac{27}{20}$$
$$= 8\frac{7}{20}$$

> $\frac{27}{20}$ is an improper fraction.
> $\frac{27}{20} = 1\frac{7}{20}$, which you add to the 7.

Subtracting fractions

To subtract fractions:

1. Convert any mixed numbers to improper fractions.

2. Find the common denominator of the fractions.

3. Change the fractions so the denominators are the same.

4. Subtract the top numbers of the two fractions.

5. Cancel if necessary.

WORKIT!

Work out $2\frac{3}{4} - 1\frac{1}{3}$.

$$2\frac{3}{4} - 1\frac{1}{3} = \frac{11}{4} - \frac{4}{3}$$
$$= \frac{33}{12} - \frac{16}{12}$$
$$= \frac{17}{12}$$
$$= 1\frac{5}{12}$$

Multiplication of fractions

To multiply two fractions such as $\frac{2}{3} \times \frac{5}{8}$:

1. Multiply the top numbers, then multiply the bottom numbers: $\frac{2}{3} \times \frac{5}{8} = \frac{10}{24}$.

2. Check whether the fraction can be cancelled further: $\frac{10}{24} = \frac{5}{12}$.

 STRETCH IT!

You can rearrange the top numbers or the bottom numbers so that the fractions cancel and the multiplication is easier: $\frac{2}{3} \times \frac{5}{8} = \frac{2}{8} \times \frac{5}{3} = \frac{1}{4} \times \frac{5}{3} = \frac{5}{12}$.

WORKIT!

Work out

a $\frac{15}{28} \times \frac{7}{20}$

$$\frac{15}{28} \times \frac{7}{20} = \frac{15}{20} \times \frac{7}{28} = \frac{3}{4} \times \frac{1}{4} = \frac{3}{16}$$

> Rearrange the top numbers and cancel to make the multiplication easier.

b $\frac{3}{4} \times \frac{2}{15}$

$$\frac{3}{4} \times \frac{2}{15} = \frac{3}{15} \times \frac{2}{4} = \frac{1}{5} \times \frac{1}{2} = \frac{1}{10}$$

> Always check that the answer cannot be cancelled further.

c $\frac{5}{14} \times 64$

$$\frac{5}{14} \times 64 = \frac{5}{14} \times \frac{64}{1} = \frac{5}{1} \times \frac{64}{14} = \frac{5}{1} \times \frac{32}{7} = \frac{160}{7} = 22\frac{6}{7}$$

Multiplication of mixed numbers

To multiply mixed numbers such as $2\frac{1}{2} \times 1\frac{2}{5}$

1. Convert any mixed numbers to improper fractions: $2\frac{1}{2} \times 1\frac{2}{5} = \frac{5}{2} \times \frac{7}{5}$.

2. Follow the steps for multiplying two fractions. Remember to simplify your answer if possible: $\frac{5}{2} \times \frac{7}{5} = \frac{35}{10} = \frac{7}{2} = 3\frac{1}{2}$

NAILIT!

Check that you can do fractions on your calculator. Most fraction questions will appear in the non-calculator paper but fractions will also appear in questions on probability. You can often use a calculator to simplify a fraction.

Division of fractions

To divide two fractions, for example $\frac{4}{5} \div \frac{2}{5}$, turn the second fraction upside down and replace the division sign with a multiplication sign.

$$\frac{4}{5} \div \frac{2}{5} = \frac{4}{5} \times \frac{5}{2} = 2$$

Division of mixed numbers

As for multiplication, first change the mixed numbers to improper fractions. Then turn the second fraction upside down and replace the division by a multiplication sign.

Ordering fractions

You may be asked to order fractions into ascending order (smallest first) or descending order (largest first). To order fractions without using a calculator, write them with a common denominator. Then they will be easy to compare.

WORKIT!

Express $3\frac{3}{4} \div \frac{5}{8}$ as a single fraction. Simplify your answer.

$$3\frac{3}{4} \div \frac{5}{8} = \frac{15}{4} \div \frac{5}{8}$$

$$= \frac{15}{4} \times \frac{8}{5}$$

$$= \frac{3}{1} \times \frac{2}{1} = 6$$

WORKIT!

Rearrange these fractions in order of size, starting from the smallest.

$$\frac{5}{6} \qquad\qquad \frac{7}{9} \qquad\qquad \frac{67}{72} \qquad\qquad \frac{2}{3}$$

$$\frac{5}{6} = \frac{60}{72} \qquad \frac{7}{9} = \frac{56}{72} \qquad \frac{67}{72} \qquad \frac{2}{3} = \frac{48}{72}$$

The order is $\frac{2}{3} \quad \frac{7}{9} \quad \frac{5}{6} \quad \frac{67}{72}$

> Write all the fractions with the same denominator.
> The common denominator for these four fractions is 72.

NAILIT!

Cancelling the fractions first will make it easier to spot the common denominator, as the denominators will be smaller.

Make sure your answer uses the original fractions and not the ones all with the same denominator.

DOIT!

Make a set of revision cards with an example of each type of calculation.

✓ CHECKIT!

1 Here are some fractions:

$$1\frac{1}{5} \qquad \frac{16}{5} \qquad \frac{3}{4} \qquad \frac{5}{8} \qquad \frac{9}{10}$$

  a Write down the fraction that is an improper fraction, and write it as a mixed number.

  b Write down the fraction that is a mixed number, and write it as an improper fraction.

  c Write them in ascending order.

  d Add the mixed number and the improper fraction.

  e Subtract the smallest number from the largest number.

2 Write down the fractions from the following list that are equivalent to the fraction $\frac{1}{3}$.

$$\frac{9}{25} \qquad \frac{15}{45} \qquad \frac{4}{12} \qquad \frac{2}{5} \qquad \frac{16}{48}$$

3 Work out the following. Give each answer as a mixed number in its simplest form.

  a $3\frac{1}{3} \times 2\frac{1}{5}$ **b** $1\frac{3}{4} \div \frac{1}{2}$

4 Ravi spends $\frac{1}{3}$ of his weekly wage on rent, $\frac{1}{5}$ on bills and $\frac{1}{4}$ on food.

  What fraction of his weekly wage does he have left?

Different types of number

Multiples, factors and prime factors

You need to remember these terms.

SNAPIT! Multiples and factors

- **Multiples** of a number are the answer you get when you multiply that number by another integer. For example, $3 \times 11 = 33$, so the number 33 is a multiple of 3. 33 is also a multiple of 11.

- **Factors (also called divisors)** are numbers that divide exactly into another number. For example, the factors of 24 are 1, 2, 3, 4, 6, 8, 12 and 24 as all these whole numbers divide exactly into 24. **Prime factors** are factors that are also prime numbers.

- **Common factors** are those factors that are shared by two different numbers. The common factors of 12 and 18 are 1, 2, 3 and 6. To find common factors, list all the factors of each number and then list the factors that are in **both** lists.

- **Highest common factor (HCF)** of two numbers is the highest factor that will divide into both numbers exactly.

- **Lowest common multiple (LCM)** of two numbers is the lowest number that is a multiple of both numbers.

WORKIT!

1 List the common factors of 100 and 32.

Factors of 100:   ① ② ④ 5, 10, 20, 25, 50, 100

Factors of 32:   ① ② ④ 8, 16, 32

Common factors: 1, 2, 4

> Circle the numbers that appear in both lists and write these common factors in a new list.

> Circle the highest number that appears in both lists.

2 Find the highest common factor of 36 and 96.

Factors of 36:   1, 2, 3, 4, 6, 9, ⑫, 18, 36

Factors of 96:   1, 2, 3, 4, 6, 8, ⑫ 16, 24, 32, 48, 96

Highest common factor of 36 and 96 is 12.

> The highest common factor is the highest number appearing in both lists.

To find the lowest common multiple, list the multiples of both numbers. For example, the multiples of 3 are 3, 6, 9, 12, 15, 18, 21, 24, 27, 30,… and the multiples of 5 are 5, 10, 15, 20, 25, 30,… The lowest common multiple of 3 and 5 is the lowest number in both lists: 15.

For larger numbers, you can use the following method to work out the lowest common multiple. List the prime factors for each number and then identify those factors that are common to both. Multiply the factors of the two numbers together, but only include the common factors once.

WORKIT!

Find the lowest common multiple of 28 and 36.

$28 = 2 \times 2 \times 7$ $36 = 2 \times 2 \times 3 \times 3$

$LCM = 2 \times 2 \times 3 \times 3 \times 7 = 252$ ◄ Notice that the 2 × 2 part appears in both so it should be included only once in the LCM.

WORKIT!

Jack is organising a barbecue. He goes to the supermarket and buys burgers in packs of 4, bread rolls in packs of 6 and sausages in packs of 8.

He buys the same number of burgers, bread rolls and sausages.

What is the minimum number of each pack he could buy? Show all your workings.

Burgers (multiples of 4)   4, 8, 12, 16, 20, (24) 28, 32, 36,...

Bread rolls (multiples of 6)   6, 12, 18, (24) 30, 36, 42, 48,...

Sausages (multiples of 8)   8, 16, (24) 32, 40, 48, 56,... ◄ You need to find the multiples of each item of food and then find the lowest number that is common to all three lists (the LCM).

The lowest common multiple for the three lists is 24, giving the minimum number of packs:

6 packs of burgers   4 packs of bread rolls   3 packs of sausages

Prime numbers

A number is **prime** if it has exactly two factors: 1 and itself. 1 is not prime, because it has only one factor. This means 2 is the smallest prime number. It is the only even prime number.

Writing a number as a product of prime factors

Every integer greater than or equal to 2 can be written as a product of prime factors.

DO IT!

There are lots of terms to remember. Produce pairs of cards for each term and its definition. Practise matching the pairs of cards regularly.

Keep breaking numbers into their factors until all the resulting numbers are prime.

WORKIT!

1 Write the number 230 as the product of prime factors.

$230 = 23 \times 10 = 23 \times 5 \times 2$

2 The number 1440 can be expressed as the product of prime factors in the form $1440 = 2^p \times 3^q \times 5^r$. Find the values of the integers p, q and r.

$1440 = 144 \times 10 = 12 \times 12 \times 5 \times 2$
$= 6 \times 2 \times 6 \times 2 \times 5 \times 2$
$= 3 \times 2 \times 2 \times 3 \times 2 \times 2 \times 5 \times 2$
$= 2^5 \times 3^2 \times 5$

$p = 5, q = 2$ and $r = 1$

✓ CHECKIT!

1 5, 7, 16.

a Which number is one more than a multiple of 5?

b Which number is one less than a factor of 12?

c Which number is **not** a prime number?

2 Write the number 300 as the product of prime factors.

3 Three grandchildren call around to see their grandmother every 12 days, 16 days and 18 days, respectively. Find out how often all three call on the same day.

4 a When written as the product of prime factors, $360 = 2^3 \times 3^2 \times 5$.

Write 756 as the product of prime factors.

b Work out the highest common factor of 360 and 756.

Listing strategies

Listing strategies help solve certain types of problems such as working out combinations of items. Once the total number of combinations has been found, it is possible to work out probabilities of certain events.

WORKIT!

A restaurant offers a fixed price menu where you order any one starter and one main course from the following lists.

Starters	Main courses
Soup	Pizza
Pâté	Lasagne
	Spaghetti
Mushrooms	Salmon

Write the first item in the starters list with each main course. Repeat for the next starter. Keep doing this until all the starters have been linked.

John likes to have a different combination of starter and main course each day.
Over how many days could he have a different combination?

Soup Pizza	Soup Lasagne	Soup Spaghetti	Soup Salmon
Pâté Pizza	Pâté Lasagne	Pâté Spaghetti	Pâté Salmon
Mushrooms Pizza	Mushrooms Lasagne	Mushrooms Spaghetti	Mushrooms Salmon

This gives 12 combinations.

John could go 12 days having different combinations.

You could be asked about probabilities. For example, 'Meals are given out at random. What is the probability that John will get soup and salmon?'
Soup and salmon is one of the 12 combinations worked out in the example above, so:

Probability of soup and salmon $= \frac{1}{12}$.

WORKIT!

There are 5 men and 4 women in a quiz team.

a One of the men and one of the women are to be chosen as a pair to go onto a TV quiz. How many possible pairs are there?

> Each of the men could be paired with any of the 4 women.

Number of pairs = 5 × 4 = 20.

b In a different quiz two men are to be chosen. How many possible pairs of two men are there?

> You might think that the first man can be chosen from 5 and the second man can be chosen from 4 making a total of 5 × 4 = 20 possible pairs. This would be wrong.

Call the men A, B, C, D and E.

If A was chosen first then B, C, D and E would be left for the second choice so this would give:

AB, AC, AD, AE

If B was chosen first then A, C, D and E could be possible second choices to give:

BA, BC, BD, BE

This process could be continued choosing C, D and then E as the first choice to give the following pairs.

CA, CB, CD, CE   DA, DB, DC, DE   EA, EB, EC, ED

This list reduces to:

> As you are looking at pairs, the order of selection does not matter: AB and BA are the same pair.
>
> Strike off those that are the same pair.

AB, AC, AD, AE   BC, BD, BE   CD, CE   DE

This is 10 pairs.

There are 10 possible pairs of two men.

NAILIT!

If you relate the pairs to combinations of people by calling them A, B, C, etc. then this question becomes easier.

CHECKIT!

1 How many different outcomes are there when tossing a coin and rolling a six-sided dice?

2 A 3-digit number is picked.

The first number is not zero.

The 3-digit number is a multiple of 5.

How many 3-digit numbers could be picked?

The order of operations in calculations

When following a series of calculations there is an order to be used. The word BIDMAS is used to help remember the order. BIDMAS stands for **B**rackets, **I**ndices (powers and roots), **D**ivision, **M**ultiplication, **A**ddition and **S**ubtraction. When you are evaluating (working out) a calculation, start with brackets, then indices, also known as orders (i.e. powers and roots), then division and multiplication, then addition and subtraction.

WORKIT!

Do the division and multiplication first, then the subtraction.

Do the calculation in the brackets first.

There are no brackets so deal with indices (i.e. powers and roots) first.

The division line acts like brackets so the addition can be considered to be in brackets and should be done first.

Work out

a $2 + 3 \times 4$

$2 + 3 \times 4 = 2 + 12 = 14$

b $5 \times 8 - 10 \div 2$

$5 \times 8 - 10 \div 2 = 40 - 5 = 35$

c $3 \times (24 - 9)$

$3 \times (24 - 9) = 3 \times 15 = 45$

d $172 - 8 \times 3^2$

$172 - 8 \times 3^2 = 172 - 8 \times 9$
$= 172 - 72 = 100$

e $\dfrac{21 + 7}{4}$

$\dfrac{21 + 7}{4} = \dfrac{28}{4} = 7$

Sometimes brackets are added for emphasis. For example these two expressions are identical.

$16 \times 3 - 24 \div 6$ $(16 \times 3) - (24 \div 6)$

WORKIT!

The calculations inside the brackets are worked out first.

The square root applies to everything under the line, so it is worked out last. Notice the sign change where two minuses are multiplied together.

Work out

a $(6 - 4)^2 + (8 \div 4)^3$ $(6 - 4)^2 + (8 \div 4)^3 = 2^2 + 2^3 = 4 + 8 = 12$

b $\sqrt{8^2 - 4 \times 3 \times (-3)}$ $\sqrt{8^2 - 4 \times 3 \times (-3)} = \sqrt{64 + 36} = \sqrt{100} = \pm 10$

 CHECKIT!

1 Work out

   a $10 + 4 \times 2$ **b** $5 \times 3 - 4 \div 2$

   c $(7 - 4)^2 + (8 \div 2)^2$

2 Work out

   a $2 + 3 \div \dfrac{1}{3} - 1$ **b** $15 - (4 - 6)^3$

   c $\sqrt{4 - 3 \times (-7)}$

Indices

Indices is another name for powers. For example, in the number 2^3, the **index** is the power to which 2 is raised (3 in this case). There are rules that you need to remember when dealing with indices.

Laws of indices

The **laws of indices** are rules that apply when the same number is raised to a power. For example, the rules would apply to $2^5 \times 2^4$ because both terms in the expression are 2 raised to a power. They would not apply to $2^3 \times 5^4$ where the two numbers are different (i.e. 2 and 5).

SNAPIT!   Laws of indices

Multiplication	$a^m \times a^n = a^{m+n}$
Division	$a^m \div a^n = a^{m-n}$
Powers of powers	$(a^m)^n = a^{m \times n} = a^{mn}$
Negative indices	$a^{-m} = \frac{1}{a^m}$ (provided $a \neq 0$)
Fractional indices	$a^{\frac{m}{n}} = \sqrt[n]{a^m} = (\sqrt[n]{a})^m$
Negative and fractional indices	$a^{-\frac{m}{n}} = \frac{1}{a^{\frac{m}{n}}} = \frac{1}{\sqrt[n]{a^m}}$ or $\frac{1}{(\sqrt[n]{a})^m}$
Zero indices	$a^0 = 1$ (provided $a \neq 0$)

NAILIT!

Remember that the final index does not always have to be positive.

NAILIT!

You need to recognise square and cube numbers.

WORKIT!

1 Express these as single powers.

a $2^3 \times 2^5$

$2^3 \times 2^5 = 2^{(3+5)} = 2^8$

> Add 3 and 5, because $2^3 \times 2^5 = 2 \times 2 \times 2 \times 2 \times 2 \times 2 \times 2 \times 2$.

b $\frac{5^5}{5^7}$

$\frac{5^5}{5^7} = 5^{(5-7)} = 5^{-2}$

> Subtract 7 from 5 because $\frac{5^5}{5^7}$
> $= \frac{5 \times 5 \times 5 \times 5 \times 5}{5 \times 5 \times 5 \times 5 \times 5 \times 5 \times 5}$
> $= \frac{1}{5^2}$.

c $(2^{-2})^3$

$(2^{-2})^3 = 2^{(-2 \times 3)} = 2^{-6}$

2 Evaluate

a 2^{-3}

$2^{-3} = \frac{1}{2^3} = \frac{1}{8} = 0.125$

> You can give your answer as a fraction or a decimal, unless the question specifies.

b $27^{\frac{1}{3}}$

$27^{\frac{1}{3}} = \sqrt[3]{27} = 3$

c $8^{\frac{2}{3}}$

$8^{\frac{2}{3}} = (\sqrt[3]{8})^2 = 2^2 = 4$

> Alternatively you could start by writing $(\sqrt{16^3})^{-1}$

d $16^{-\frac{3}{2}}$

$16^{-\frac{3}{2}} = \dfrac{1}{16^{\frac{3}{2}}} = \dfrac{1}{(\sqrt{16})^3} = \dfrac{1}{\pm 4^3} = \pm \dfrac{1}{64}$

> Find the square root of 16 and then cube the answer, rather than cubing 16 first and then finding the square root.

e 5^0

$5^0 = 1$

> Any number to the power 0 is equal to 1.

Simplifying more complex expressions

You can use the rules of indices with more complex expressions, for example expressions containing more than two terms, or expressions including powers of two different numbers.

WORKIT!

Express $\dfrac{2^{-2} \times 2^5}{2}$ as a single power of 2.

$\dfrac{2^{-2} \times 2^5}{2} = \dfrac{2^{-2} \times 2^5}{2^1} = 2^{-2+5-1} = 2^2$

> $2 = 2^1$

If you have an expression with powers of two different numbers, you can combine the powers of each number separately.

WORKIT!

Simplify $(3^5)^2 \times 5^2 \times 3^{-4} \times 5$.

$(3^5)^2 \times 5^2 \times 3^{-4} \times 5 = 3^{(5 \times 2 - 4)} \times 5^{(2 + 1)}$

> Gather together the terms for each number and simplify.

$= 3^6 \times 5^3$

✓ CHECKIT!

1 Express each of the following as a single power.

a $7^7 \times 7^3$

b $3^{-2} \div 3^4$

c $(5^4)^5$

2 Simplify

a $\dfrac{5^4 \times 5^6}{5^3}$

b $(6^{-4} \times 6^2)^3 \div 6^{-3}$

c $5^{-7} \times 2^8 \times 5^4 \times 2^2$

d $\dfrac{7^4 \times 11^3 \times (7^{-2})^{-3}}{11^4}$

3 Find the value of

a 16^0

b $100^{\frac{1}{2}}$

c $64^{\frac{2}{3}}$

d $25^{-\frac{1}{2}}$

e $27^{\frac{1}{3}} \times 36^{\frac{1}{2}}$

4 Find the value of x such that $(3^{2x})^2 = 81$.

Surds

A surd is the square root of a non-square number. Surds are irrational numbers. This means that they cannot be expressed as fractions, recurring decimals or terminating decimals. For example, $\sqrt{5}$ is a surd.

Surds can be simplified like this:

$$\sqrt{18} = \sqrt{9 \times 2} = 3\sqrt{2}$$

> The number 18 is broken down into two factors, one of which is a square number. 9 is a perfect square so its square root is a whole number.

WORKIT!

Write $\sqrt{50}$ in the form $k\sqrt{2}$, where k is an integer.

$\sqrt{50} = \sqrt{25 \times 2} = 5\sqrt{2}$

STRETCHIT!

Always try to find the largest square factor. For example, $\sqrt{80}$ could be written as $\sqrt{4 \times 20}$, or as $\sqrt{16 \times 5}$ but both expressions need further simplification. It is quicker to spot that 16 is the highest square factor of 80.

Simplifying surds

Here are some general rules when manipulating surds.

SNAPIT!   Simplifying surds

$$\sqrt{a} \times \sqrt{a} = a$$
$$\sqrt{a} \times \sqrt{b} = \sqrt{ab}$$
$$(\sqrt{a} + \sqrt{b})(\sqrt{a} - \sqrt{b}) = a - b$$
$$(\sqrt{a} + \sqrt{b})(\sqrt{a} + \sqrt{b}) = a + \sqrt{ab} + \sqrt{ab} + b = a + 2\sqrt{ab} + b$$
$$(a + \sqrt{b})(c + \sqrt{d}) = ac + a\sqrt{d} + c\sqrt{b} + \sqrt{bd}$$
$$a\sqrt{b} \times c\sqrt{b} = abc$$
$$a\sqrt{b} + c\sqrt{b} = (a + c)\sqrt{b}$$

These examples show how the above rules can be applied:

1. $(\sqrt{3})^2 = \sqrt{3} \times \sqrt{3} = 3$

2. $(5\sqrt{2})^2 = 5\sqrt{2} \times 5\sqrt{2} = 25 \times 2 = 50$

3. $3\sqrt{2} \times 4\sqrt{2} = 12 \times 2 = 24$

4. $3\sqrt{2} + 2\sqrt{2} = 5\sqrt{2}$

5. $(2 + \sqrt{7})(2 + \sqrt{7}) = 4 + 2\sqrt{7} + 2\sqrt{7} + 7$
$$= 11 + 4\sqrt{7}$$

> Use the normal rules for multiplying out pairs of brackets.

6. $(1 + \sqrt{3})(5 - \sqrt{12}) = 1(5 - \sqrt{12}) + \sqrt{3}(5 - \sqrt{12})$
$$= 5 - 2\sqrt{3} + 5\sqrt{3} - \sqrt{36}$$
$$= -1 + 3\sqrt{3}$$

Rationalising surds

If you have a fraction with a surd on the bottom the surd needs to be avoided in your final answer (i.e. **rationalised**). This is done by multiplying the numerator and denominator of the fraction by the surd. Rationalising makes sure that the denominator is no longer an irrational number.

$$\frac{1}{\sqrt{3}} = \frac{1}{\sqrt{3}} \times \frac{\sqrt{3}}{\sqrt{3}} = \frac{\sqrt{3}}{3}$$ ← The fraction is simplified when there are no surds in the denominator.

If the denominator has the form $a + \sqrt{b}$, for example $\frac{1}{1-\sqrt{2}}$, remove the irrational number in the denominator by multiplying both the numerator and denominator by the expression that is the same as the denominator but with the opposite sign for the surd, in this case $1 + \sqrt{2}$.

$$\frac{1}{1-\sqrt{2}} = \frac{1}{(1-\sqrt{2})} \times \frac{(1+\sqrt{2})}{(1+\sqrt{2})} = \frac{1+\sqrt{2}}{1-2} = \frac{1+\sqrt{2}}{-1} = -1 - \sqrt{2}$$

WORKIT!

Rationalise the denominator (i.e. eliminate the surd from the denominator) and simplify the result.

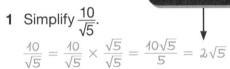

1 Simplify $\frac{10}{\sqrt{5}}$.

$$\frac{10}{\sqrt{5}} = \frac{10}{\sqrt{5}} \times \frac{\sqrt{5}}{\sqrt{5}} = \frac{10\sqrt{5}}{5} = 2\sqrt{5}$$

2 Simplify $\frac{1}{2-\sqrt{5}}$.

Rationalise the denominator by multiplying the numerator and denominator by the conjugate of the denominator $(2 + \sqrt{5})$.

$$\frac{1}{(2-\sqrt{5})} \frac{(2+\sqrt{5})}{(2+\sqrt{5})} = \frac{2+\sqrt{5}}{4-5} = -2 - \sqrt{5}$$

3 Show that $\sqrt{3}\left(5\sqrt{12} + \sqrt{3}\right)$ can be simplified to give 33.

$$\sqrt{3}\left(5\sqrt{12} + \sqrt{3}\right) = 5\sqrt{36} + 3 = 5 \times 6 + 3 = 33$$

NAILIT!

'Show that' means show working that leads to the answer that is given in the question. In this case, show working that leads to the answer 33.

CHECKIT!

1 Simplify

 a $\sqrt{3} \times \sqrt{2}$

 b $(\sqrt{5})^2$

 c $2\sqrt{3} \times 3\sqrt{3}$

 d $(2\sqrt{5})^2$

2 $\sqrt{28}$ can be written as $a\sqrt{7}$.
 Find the value of a.

3 Express as a multiple of the smallest possible surd

 a $\sqrt{45}$

 b $\sqrt{72}$

4 Simplify

 a $\frac{16}{3\sqrt{2}}$

 b $\frac{18}{4+\sqrt{7}}$

5 Simplify fully

 a $(1 + \sqrt{5})(1 - \sqrt{5})$

 b $(2 + \sqrt{3})^2$

 c $(1 + \sqrt{3})(2 + \sqrt{3})$

Standard form

Representing numbers in standard form

Standard form is used as an alternative way of writing very large or very small numbers.

Numbers in standard form are represented as:

$a \times 10^n$

where $1 \leq a < 10$ (between 1 and 10, including 1 but not including 10)

n is an integer (a whole number, positive or negative or zero).

		Valid values	Invalid values
a		1, 1.2, 3.555, 9.8009	0.8, 0.25, 0.999, 10, 11.4, 200
n		1, 5, 10, 23, 0, −1, −5, −10	1.5, 0.25, −0.02, −1.5

WORKIT!

State whether each of the following numbers is in standard form. If they are not in standard form, explain why not.

a 1.00×10^3

Yes

b 0.75×10^{-2}

No, a is less than 1.

c 3.667×10^{-5}

Yes

d $7.5 \times 10^{2.5}$

No, n is not an integer.

e 10×10^{-100}

No, a must be less than 10.

> Look at the values of a and n in $a \times 10^n$.

Converting large numbers into standard form

To convert large ordinary numbers, such as 4500, into standard form:

1. Write the digits of the number in a place value table so that it is a number between 1 and 10.

2. Work out how the digits move to make the original number: three places to the left, so multiply by 10^3.

> To multiply, move digits left.

1000	100	10	1	$\frac{1}{10}$	$\frac{1}{100}$	$\frac{1}{1000}$
			4	5		

$4.5 \times 10^3 = $

1000	100	10	1	$\frac{1}{10}$	$\frac{1}{100}$	$\frac{1}{1000}$
4	5	0	0			

WORKIT!

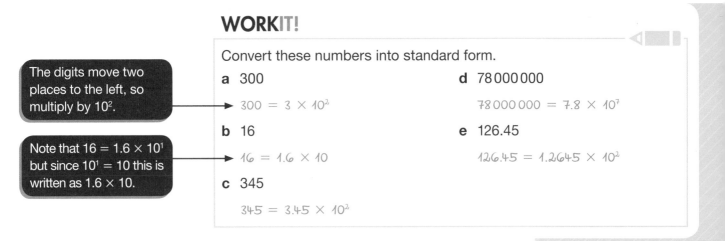

Convert these numbers into standard form.

a 300

$300 = 3 \times 10^2$

b 16

$16 = 1.6 \times 10$

c 345

$345 = 3.45 \times 10^2$

d 78 000 000

$78\,000\,000 = 7.8 \times 10^7$

e 126.45

$126.45 = 1.2645 \times 10^2$

The digits move two places to the left, so multiply by 10^2.

Note that $16 = 1.6 \times 10^1$ but since $10^1 = 10$ this is written as 1.6×10.

Converting small numbers into standard form

To convert small ordinary numbers, such as 0.0003, into standard form:

1. Write the digits of the number in a place value table so that it is a number between 1 and 10.

2. Work out how the digits move to make the original number: four places to the right, so divide by 10^4. However, in standard form you must multiply by 10^n. Dividing by 10^4 is the same as multiplying by 10^{-4}.

This is one of the laws of indices covered on page 26.

1	$\frac{1}{10}$	$\frac{1}{100}$	$\frac{1}{1000}$	$\frac{1}{10\,000}$
3				
0	0	0	0	3

$3 \div 10^4 = 3 \times 10^{-4} = $

To divide, move digits right.

WORKIT!

Convert these numbers into standard form.

a 0.008

$0.008 = 8 \times 10^{-3}$

b 0.0000357

$0.0000357 = 3.57 \times 10^{-5}$

The digits move three places to the right, so multiply by 10^{-3}.

c 0.00000000045

$0.00000000045 = 4.5 \times 10^{-10}$

NAILIT!

Always check that your final answer includes a number between 1 and 10 multiplied by 10 to a power.

Converting other numbers into standard form

Sometimes a number looks like it is in standard form but isn't. For example, 12×10^{-4} looks like it is in standard form because it contains a power of 10. However, the number 12 is not between 1 and 10.

To convert numbers like this into standard form:

1. Convert the number 12 into a valid value for a: $12 = 1.2 \times 10$.

2. Multiply the original power of 10 by this number:
$12 \times 10^{-4} = 1.2 \times 10 \times 10^{-4} = 1.2 \times 10^{-3}$.

NAILIT!

Check that you know how to enter standard form numbers on your calculator, and how to interpret standard form answers. The key on your calculator may be EXP, ×10ˣ or EE.

DO IT!

Use the internet to look up some very large and very small sizes. Convert these to standard form.

WORKIT!

Write these numbers in standard form.

a 125×10^5

$$125 \times 10^5 = 1.25 \times 10^2 \times 10^5$$
$$= 1.25 \times 10^7$$

b 250×10^{-3}

$$250 \times 10^{-3} = 2.5 \times 10^2 \times 10^{-3}$$
$$= 2.5 \times 10^{-1}$$

Calculations using standard form

Here are some calculations that can be carried out without using a calculator.

WORKIT!

Multiply the ordinary numbers first: 3×4. Then add the indices together to give the power of 10.

The answer is not in standard form. Write 12 in standard form: $12 = 1.2 \times 10$ and then multiply by the power of 10.

Work out these calculations without using a calculator. Give your answers in standard form.

a $3 \times 10^5 \times 4 \times 10^2$

$$3 \times 10^5 \times 4 \times 10^2 = 12 \times 10^7$$
$$= 1.2 \times 10 \times 10^7$$
$$= 1.2 \times 10^8$$

b $4 \times 10^{-8} \times 4 \times 10^{-2}$

$$4 \times 10^{-8} \times 4 \times 10^{-2}$$
$$= 16 \times 10^{-10} = 1.6 \times 10^{-9}$$

c $\dfrac{8 \times 10^{-3}}{2 \times 10^{-2}}$

$$\frac{8 \times 10^{-3}}{2 \times 10^{-2}} = 4 \times 10^{-3-(-2)}$$
$$= 4 \times 10^{-1}$$

16×10^{-10} needs changing to standard form.

CHECK IT!

1 Write as ordinary numbers

   a 5×10^{-3} **b** 5.65×10^5

2 Express in standard form

   a $25\,000$ **c** 0.05×10^4

   b 0.00125 **d** 14×10^{-3}

3 $a = 8 \times 10^4$, $b = 3 \times 10^{-2}$ and $c = 4 \times 10^2$
Calculate the following, giving your answers in standard form.

   a b^2 **c** $\dfrac{a}{c}$

   b ab **d** $a + c$

4 a In 2016 the richest person in Britain had a wealth of £13.3 billion.

Write this amount in standard form. (1 billion is one thousand million.)

   b He decides to give his money away by giving £500 to as many people as he can. How many people could he give £500 to? Give your answer as an ordinary number.

5 a Write $15\,500$ in standard form.

   b Write 6.55×10^5 as an ordinary number.

   c Work out $\dfrac{1.25 \times 10^5}{2.5 \times 10^2}$, giving your answer in standard form.

   d Evaluate $\sqrt{1.6 \times 10^7}$, giving your answer in standard form.

Converting between fractions and decimals

Here are some familiar fractions and their decimal equivalents.

 SNAP **Converting between fractions and decimals**

$\frac{1}{2} = 0.5$ $\frac{1}{4} = 0.25$ $\frac{1}{8} = 0.125$ $\frac{1}{10} = 0.1$

$\frac{1}{20} = 0.05$ $\frac{1}{3} = 0.333...$ $\frac{2}{3} = 0.666...$

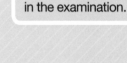 **NAIL**IT!

You should remember all these conversions as they often come up in the examination.

To convert a fraction, for example $\frac{4}{25}$, to a decimal, divide the bottom number (25) into the top number (4).

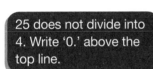 25 does not divide into 4. Write '0.' above the top line.

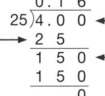

$$\begin{array}{r} 0.1\ 6 \\ 25\overline{)4.0\ 0} \\ 2\ 5 \\ \hline 1\ 5\ 0 \\ 1\ 5\ 0 \\ \hline 0 \end{array}$$

25 divides into 40 once, with remainder 15. Write '1' above the top line.

Bring the zero down from the number you are dividing.

 **NAIL**IT!

Use long division in a non-calculator paper and your calculator in a calculator paper.

Terminating and recurring decimals

A **terminating** decimal has digits which do not go on forever, for example 0.25 and 1.56. In a **recurring** decimal, a number or pattern of numbers after the decimal point repeats. For example, $\frac{1}{3} = 0.333333...$, $\frac{7}{22} = 0.318181818...$ and $\frac{8}{27} = 0.296296296...$ are all recurring decimals.

To decide whether a fraction will produce a terminating or recurring decimal:

1 Cancel the fraction as far as possible.

2 Write the denominator as a product of prime factors.

3 If the prime factors in the denominator are only 2 and/or 5, then the decimal will terminate.

Converting terminating decimals to fractions

To convert a terminating decimal such as 0.625 into a fraction:

1 Count the number of digits after the decimal point.

2 Write the decimal in a place value table:

1		$\frac{1}{10}$	$\frac{1}{100}$	$\frac{1}{1000}$	$\frac{1}{10\,000}$
0		6	2	5	

③ Use the final digit to work out the denominator: final digit (5) is thousandths, so the denominator is 1000, giving $\frac{625}{1000}$.

④ Cancel the fraction as far as possible: $\frac{625}{1000} = \frac{5}{8}$.

Changing recurring decimals to fractions

Recurring decimals can be written with a dot over the number if a single number repeats (e.g. $\frac{1}{3}$ 0.3333333… = 0.$\dot{3}$) or dots over the first and the last number of a block of repeating numbers (e.g. $\frac{7}{22}$ 0.318181818… = 0.3$\dot{1}\dot{8}$).

> Multiply by multiples of 10 until you have two numbers with exactly the same digits after the decimal point.

WORKIT!

Write the recurring decimal 0.8$\dot{6}$ as a fraction in its simplest form.

Let $x = 0.8\dot{6} = 0.86666666…$

$10x = 8.6666666…$

$100x = 86.6666666…$

$100x - 10x = 86.6666666…$
$\qquad\qquad\quad - 8.6666666…$

$90x = 78$

$x = \frac{78}{90}$

$x = \frac{13}{15}$

> Cancel the fraction by dividing both top and bottom by 6.

> Subtract 10x from 100x to eliminate the digits after the decimal point.

WORKIT!

Write the recurring decimal 0.$\dot{3}\dot{9}$ as a fraction in its simplest form.

Let $x = 0.\dot{3}\dot{9} = 0.39393939…$

$100x = 39.39393939…$

$100x - x = 39.39393939…$
$\qquad\qquad\; - 0.39393939…$

$99x = 39$

$x = \frac{39}{99} = \frac{13}{33}$

> Subtract x from 100x to eliminate the digits after the decimal point.

DOIT!

Make cards of the numbers 1–20. Shuffle the cards and pick two to make a fraction then work out if the fraction is recurring or terminal.

NAILIT!

If one of these questions appears in the calculator paper, divide the fraction to check you get back to the original recurring decimal.

CHECKIT!

1 Express these fractions as decimals.

a $\frac{43}{100}$ b $\frac{3}{8}$ c $\frac{11}{20}$

2 Express these terminating decimals as fractions in their simplest form.

a 0.8 b 0.45 c 0.584

3 Express these recurring decimals as fractions in their simplest form.

a 0.$\dot{7}$ b 0.0$\dot{4}$ c 0.9$\dot{5}\dot{4}$

4 Express these decimals as fractions in their simplest form.

a 0.$\dot{5}1\dot{8}$ b 0.76

5 Show that 0.$\dot{7}$ + $\frac{2}{9}$ = 1.

Converting between fractions and percentages

A **percentage** is a fraction where the denominator is 100. For example, $\frac{6}{100} = 6\%$.

Percentages allow values to be compared, as they are always out of 100.

Converting a percentage to a fraction

To convert a percentage to a fraction, divide the percentage by 100 and then simplify the fraction if possible.

WORKIT!

Convert 45% to a fraction in its simplest form.

$45\% = \frac{45}{100} = \frac{9}{20}$

NAILIT!

Failing to cancel fractions will lose you marks. Always look at the fraction carefully to see if it can be cancelled.

Converting a fraction to a percentage

If possible, multiply the top and bottom of the fraction so that the bottom number is 100. The top number is then the percentage.

For example, $\frac{2}{5} = \frac{2 \times 20}{5 \times 20} = \frac{40}{100} = 40\%$.

Alternatively, convert the fraction to a decimal and then multiply the decimal by 100.

For example, $\frac{1}{6} = 0.16666\ldots = 16.67\%$.

NAILIT!

This method works when the denominator is a factor of 100.

WORKIT!

Convert $\frac{36}{55}$ to a percentage. Give your answer to two decimal places.

$\frac{36}{55} \times 100 = 65.45\%$ ◀——

> You can also multiply the fraction by 100 and then convert to a decimal.

DOIT!

Create a table of conversions between common fractions and percentages.

✓ CHECKIT!

1 Convert these percentages to fractions.

   a 25%   **b** 85%   **c** 68%

2 Charlie scored $\frac{65}{80}$ in a maths test and 80% in an English test. Which test did he do better in? Show your working.

3 Convert these fractions to percentages. Where necessary, give your answer to 1 decimal place.

   a $\frac{3}{10}$ **b** $\frac{4}{25}$ **c** $\frac{3}{7}$

Fractions and percentages as operators

Fractions and percentages are often used as **multipliers** when solving problems.

Replace 'of' with a multiplication sign.

WORKIT!

Work out

a 22% of £600

$$22\% \text{ of } £600 = \frac{22}{100} \times 600 = 22 \times 6 = £132$$

b $\frac{9}{20}$ of £600

$$\frac{9}{20} \text{ of } £600 = \frac{9}{20} \times 600 = 9 \times 30 = £270$$

WORKIT!

In school A, 48% of the 600 students are boys.

In school B, $\frac{4}{9}$ of the 630 students are boys.

Which school has the greater number of boys and by how many?

$$48\% \text{ of } 600 = \frac{48}{100} \times 600 = 48 \times 6 = 288$$

$$\frac{4}{9} \text{ of } 630 = \frac{4}{9} \times 630 = 4 \times 70 = 280$$

School A has the greater number of boys by 8.

Find 48% of 600 and $\frac{4}{9}$ of 630. Remember that 'of' means multiply.

WORKIT!

Joshua spends $\frac{1}{3}$ of his weekly wage on rent, $\frac{1}{4}$ on food and $\frac{1}{5}$ on entertainment. How much money does he have left each week if he earns £480 per week?

Rent: $\frac{1}{3} \times 480 = £160$ Food: $\frac{1}{4} \times 480 = £120$

Entertainment: $\frac{1}{5} \times 480 = £96$

Amount remaining $= 480 - (160 + 120 + 96) = £104$

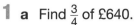

CHECKIT!

1 a Find $\frac{3}{4}$ of £640.

   b Find 15% of £30.

   c Find 95% of 80 kg.

2 In School A 56% of the pupils are boys and in School B 35% of the pupils are girls. There are 600 pupils in School A and 700 pupils in School B.

How many boys are there in each school?

Standard measurement units

Many quantities can be expressed in different units. For example, time can be expressed in seconds, minutes, hours, days or years and distance can be expressed in metres or kilometres. It is very important to be able to convert quantities to different units.

Metric unit conversions

SNAP IT!   Metric unit conversions

1 cm = 10 mm	1 tonne = 1000 kg
1 m = 100 cm	1 cm³ = 1 ml
1 km = 1000 m	1 litre = 1000 cm³
1 kg = 1000 g	1 litre = 1000 ml

WORKIT!

Convert

a 3.3 litres into ml

　　3.3 litres = 3.3 × 1000 = 3300 ml ◄──

> To convert from a larger unit to a smaller unit, always multiply.

b 850 g into kg

　　850 g = 850 ÷ 1000 = 0.85 kg ◄──

> To convert from a smaller unit to a larger unit, always divide.

c 2 m into mm.

　　2 m = 2 × 100 = 200 cm

　　2 m = 200 cm = 200 × 10 = 2000 mm ◄──

> First convert m to cm by multiplying by 100. Then multiply by 10 to convert cm to mm.

NAIL IT!

You can also convert straight from m to mm, by multiplying by 1000.

Time and money

SNAP IT!   Time

60 seconds = 1 minute	7 days = 1 week
60 minutes = 1 hour	52 weeks = 1 year
24 hours = 1 day	

Money is usually either in pounds or pence.

WORKIT!

1 A large tin of beans costs 67p.

Tins of beans are packed into trays of 24 tins. There are 20 trays on a pallet and 40 pallets are loaded onto a van.

Find the value of all the beans loaded onto the van, giving your answer in pounds.

> Convert the 67p to pounds by dividing by 100.

One tin of beans costs £0.67.

Cost of beans on van = 0.67 × 24 × 20 × 40 = £12 864

2 Ahmed earns a salary of £28 000 per year for a 40-hour week.

a Work out his monthly wage.

$$\text{Monthly wage} = \frac{\text{Yearly wage}}{12} = \frac{28\,000}{12} = £2333.33$$

b Calculate his daily rate for the month of June. Give your answer correct to the nearest pence.

> Assume he gets the same amount each month.

$$\text{Daily rate for June} = \frac{\text{Monthly wage}}{30} = \frac{2333.33}{30} = £77.78$$

NAILIT!

Never write two units next to a number, for example £2.35p. It will be marked incorrect, as it is not clear whether it is £2.35 or 2.35p.

✓ CHECKIT!

1 Convert

  a 9.7 kg to grams

  b 850 cm³ to litres

  c 2.05 km to centimetres

2 Work out how many seconds there are in one day. Write your answer in standard form.

3 A café sells a carton of juice for 34p.

It has 12 boxes, each containing 20 cartons.

Work out the value of the café's stock of juice. Give your answer in pounds.

Rounding numbers

Rounding numbers

Rather than give exact values of numbers, we often approximate them. In calculations you will be given values that need to be put into an equation to calculate an unknown value. You may be asked to give the answers to a certain number of **decimal places** or **significant figures**.

Decimal places

123.764 is given to 3 decimal places (d.p.) as there are three numbers to the right of the decimal point.

To give a number to a specified number of decimal places, look at the digit in the next decimal place. For example to round to 2 decimal places, look at the digit in the 3rd place after the decimal point:

- If this is less than 5, then the digit in the 2nd place stays the same.
- If it is 5 or greater, the digit in the 2nd place increases by 1.

WORKIT!

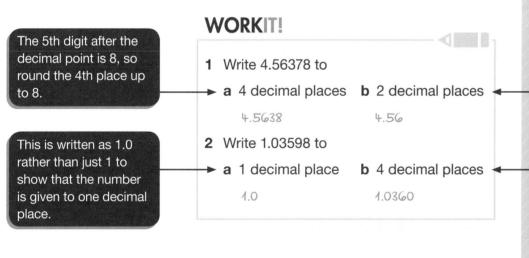

The 5th digit after the decimal point is 8, so round the 4th place up to 8.

This is written as 1.0 rather than just 1 to show that the number is given to one decimal place.

1   Write 4.56378 to

   a 4 decimal places   b 2 decimal places

     4.5638   4.56

2   Write 1.03598 to

   a 1 decimal place   b 4 decimal places

     1.0   1.0360

The 3rd digit after the decimal point is 3, so the 2nd place digit stays the same.

The 5th digit after the decimal place is 8, so the 9 in the 4th place needs to increase. So the 9 becomes 0 and the next digit to the left increases by 1.

Significant figures

Another way of rounding a number is to express it as significant figures (s.f.). This means a specified number of digits in a row (i.e. successive place values).

SNAP IT! Significant figures

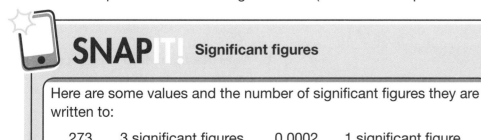

Here are some values and the number of significant figures they are written to:

273	3 significant figures	0.0002	1 significant figure
4200	2 significant figures	0.076	2 significant figures
8.31	3 significant figures	0.1005	4 significant figures

The counting for the figures starts at the first non-zero digit.

Zeros at the beginning and end of numbers are not usually significant: 4200 is to 2 s.f. and 0.0002 is to 1 s.f. However, in the middle of numbers zeros are significant as they tell you the place value for other digits: 0.1005 is to 4 s.f.

WORKIT!

Give these numbers to the stated number of significant figures.

Count the numbers, starting with the first non-zero digit. Once you reach the required significant figure, use the next digit to round up or down.

a 3.5609   2 s.f.   3.6

b 94 236   3 s.f.   94 200

Fill any place values to the right of the last significant digit with zeros, so the number still has the same number of digits.

c 0.0040658 3 s.f.   0.00407

Starting counting from the first non-zero digit, the first 3 significant figures are 406.

NAILIT!

Always read the question to see whether the answer should be given to a certain number of significant figures or decimal places or as a whole number.

DOIT!

Draw a flowchart to round to a certain number of decimal places.

CHECKIT!

1 Write each number to 3 significant figures.

   a 1259

   b 14.919

   c 0.0003079

   d 9 084 097

   e 1.8099×10^{-4}

2 Write each number to the number of decimal places specified in brackets.

   a 10.565 (1)

   b 123.9765 (3)

   c 0.02557 (2)

   d 3.9707 (3)

   e 0.00195 (3)

   f 4.098 (2)

3 Here is a number, 1989.

   a Write this number to 1 significant figure.

   b Write this number to 2 significant figures.

   c Write this number to 3 significant figures.

4 Write the number 3.755×10^{-4} as an ordinary number to 4 decimal places.

Estimation

Estimating answers

It is easy to make a mistake when entering numbers into a calculator, so it is important to have an idea what the answer should be. This means that you can recognise a wrong answer. Estimating is an important part of checking your answers.

When making estimations, write all the values in the calculation to 1 significant figure and give the answer to 1 significant figure.

WORKIT!

Estimate the answers to these calculations, giving your answers to 1 significant figure. Compare your estimates with the exact answers using your calculator.

a $2.8 \times 0.9 \times 3.4$

$2.8 \times 0.9 \times 3.4 \approx 3 \times 1 \times 3 = 9$

Working this out accurately on a calculator, $2.8 \times 0.9 \times 3.4 = 8.568$
$= 9$ (to 1 s.f.)

> Write all the numbers to 1 significant figure. The symbol $\approx$ means approximately equal to.

b $\dfrac{2.68 \times 7.89 \times 12.00}{2.98 \times 6.07 \times 11.67}$

$\dfrac{2.68 \times 7.89 \times 12.00}{2.98 \times 6.07 \times 11.67} \approx \dfrac{3 \times 8 \times 10}{3 \times 6 \times 10} = \dfrac{4}{3} = 1.\dot{3} = 1$ (to 1 s.f.)

Working this out accurately on a calculator, $\dfrac{2.68 \times 7.89 \times 12.00}{2.98 \times 6.07 \times 11.67}$
$= 1.2020 = 1$ (to 1 s.f.)

> Cancel the numbers in the numerator with those in the denominator as far as possible.

c $\sqrt{2.5 \times 4.1 \times 2.9}$

$\sqrt{2.5 \times 4.1 \times 2.9} \approx \sqrt{3 \times 4 \times 3} = \sqrt{36} = 6$ or -6 (1 s.f.)

Working this out accurately on a calculator, $\sqrt{2.5 \times 4.1 \times 2.9} = \pm 5.4521$
$= \pm 5$ (to 1 s.f.)

> Note that this is slightly different from the estimate.

Estimating square roots and powers

To estimate the value of $\sqrt{32}$ to 1 decimal place, start by finding perfect squares that are above and below the number inside the square root: $5^2 = 25$ and $6^2 = 36$, so the answer is between 5 and 6. Then decide whether 32 is nearer to 36 or 25. It's nearer to 36, so the answer is probably nearer to 6. A reasonable estimate might be 5.7.

> $\sqrt{32}$ is 5.656854249, so the estimate was pretty good.

Powers can be estimated in a similar way. To estimate $(3.6)^3$, work out the whole number powers on either side: $3^3 = 27$ and $4^3 = 64$. The value of $(3.6)^3$ will be a little over midway between them, so estimate 45.

> $(3.6)^3$ is 46.656.

WORKIT!

Estimate values for

a $\sqrt{90}$ to 1 decimal place

$9^2 = 81$ and $10^2 = 100$ so $\sqrt{90}$ lies between 9 and 10.

90 lies almost midway between 81 and 100, so 9.5 is a reasonable estimate.

$\sqrt{90} \approx 9.5$ (to 1 d.p.) ◄——

> The actual value is 9.487 (to 3 d.p.).

b $(5.25)^2$ to the nearest whole number

$5^2 = 25$ and $6^2 = 36$

5.25 is closer to 5, so estimate 28

$(5.25)^2 \approx 28$ (to nearest whole number) ◄——

> The actual value is 27.5625.

c $5^{2.8}$ to the nearest whole number.

$5^2 = 25$ and $5^3 = 125$

2.8 is closer to 3, so estimate

$5^{2.8} \approx 95$ (to nearest whole number) ◄——

> The actual value is 90.597 (to 3 d.p.)

> Allow for the fact that as the power gets larger, the values are further apart. So $5^{2.5}$ is closer to 5^2 than to 5^3.

CHECKIT!

1 Work out an estimate for

a $\dfrac{5.9 \times 189}{0.54}$

b $\sqrt{4.65} + 28.9 \div 6$

2 Use estimation to find which of the answers A, B, or C is closest to the exact answer. Approximate each number in the calculation to one significant figure.

	Calculation	Answer A	Answer B	Answer C
a	$3.45 \times 2.78 \times 0.09$	0.8	1	0.08
b	$12.56 \times 1.87 \times 0.45$	6	26	13
c	$120 \div 0.45$	300	200	60
d	$0.01 \times 0.15 \times 109$	0.3	0.2	2
e	$0.12 \times 300 \times 0.53$	15	41	9
f	$6.07 \times 3.67 \times 0.1$	4	2	5
g	$20.75 \div 6.98$	0.3	4	3
h	$0.01 \times 145 \times 35$	40	400	450
i	$6.5 \times 0.3 \times 0.01$	0.01	0.02	0.2
j	$65 \div 1050$	0.07	0.1	0.001

3 Estimate values for

a $\sqrt{45}$ to 1 decimal place

b $\sqrt{104}$ to 1 decimal place

c $(2.3)^3$

d $3^{1.4}$

Upper and lower bounds

Finding upper and lower bounds

If you measured the length of a line to the nearest mm using a ruler, the measurement could be up to 0.5 mm larger or smaller. The maximum value a measurement can have is called the **upper bound** and the minimum value is called the **lower bound**.

For a measured length of 45 mm:

upper bound = 45 + 0.5 = 45.5 mm

lower bound = 45 − 0.5 = 44.5 mm

Find half of the given unit (i.e. 1 g). Add it to the value for the upper bound and subtract it for the lower bound.

WORKIT!

The mass of coffee in a packet is 227 g to the nearest g. Find the upper and lower bounds.

Upper bound = 227 + 0.5 = 227.5 g

Lower bound = 227 − 0.5 = 226.5 g

Error intervals

The **error interval** is the range of values (between the upper and lower bounds) in which the precise value could lie. It is normally written as:

lower bound $\leq x <$ upper bound

Truncation occurs when all the digits past a certain point are cut off.

For example, the fraction $\frac{2}{3}$ = 0.66666... truncated to 2 decimal places becomes 0.66.

WORKIT!

A length, l cm, is recorded as 2.54 cm.

a Write the error interval for l where the measurement has been truncated to 3 significant figures.

Upper bound = 2.549999... cm

Lower bound = 2.54 cm

The error interval is 2.54 $\leq l <$ 2.55 cm

2.549999... is just less than 2.55.

b Write the error interval for l where the measurement has been rounded to 2 decimal places.

$\frac{0.01}{2}$ = 0.005 cm

Lower bound = 2.54 − 0.005 = 2.535 cm

Upper bound = 2.54 + 0.005 = 2.545 cm

The error interval is 2.535 $\leq l <$ 2.545 cm

This is half the smallest unit in the given measurement.

DOIT!

Draw a poster to show how to find the upper and lower bounds of the volume of a cuboid.

Calculations using upper and lower bounds

WORKIT!

$T = \frac{\sqrt{l}}{g}$ $l = 2.658$ correct to 4 significant figures

$g = 9.81$ correct to 3 significant figures

a By considering the upper bounds and the lower bounds for l and g, work out the error interval for T. Give a reason for your answer.

l has an upper bound of 2.6585 and a lower bound of 2.6575

g has an upper bound of 9.815 and a lower bound of 9.805

Maximum value of $T = \frac{\sqrt{l}}{g} = \frac{\sqrt{2.6585}}{9.805} = 0.16629$ ◀—

> The maximum value of T will be when the upper bound is used for l and the lower bound for g.

Minimum value of $T = \frac{\sqrt{l}}{g} = \frac{\sqrt{2.6575}}{9.815} = 0.16609$ ◀—

> The minimum value of T will be when the lower bound is used for l and the upper bound for g.

$0.16609 \leq T < 0.16629$

b Give a value for T to a suitable degree of accuracy.

The upper and lower bounds for T are the same when written to 3 significant figures.

So $T = 0.166$ to 3 significant figures

> To find the upper and lower bounds of values calculated from two numbers, you need to use the bounds shown in the table.

> The smallest value of $A + B$ will be made from the smallest values of both A and B.

SNAPIT! Upper and lower bounds

Operation	Calculation	Lower bound of calculation		Upper bound of calculation	
		A	B	A	B
Addition	$A + B$	Lower	Lower	Upper	Upper
Subtraction	$A - B$	Lower	Upper	Upper	Lower
Multiplication	$A \times B$	Lower	Lower	Upper	Upper
Division	$A \div B$	Lower	Upper	Upper	Lower

CHECKIT!

1 The length of a piece of wood is 145 cm to the nearest centimetre.

  a Write down the lower bound for the length of the piece of wood.

b Write down the upper bound for the length of the piece of wood.

c Express these bounds as an error interval for the length of wood, l.

1 a Using a calculator, evaluate $5\sqrt{2.43 \times 10^7}$.

b Using a calculator, evaluate $10^{\frac{4}{3}}$. Give your answer to 1 decimal place.

2 Without using a calculator, work out

a $8 - 3 \div 2$

b $(-1)^3 + 32 \div 8 + 1$

c $\frac{27}{\frac{1}{3}}$ **d** 1.65×3.6

3 Without using a calculator, simplify the following fractions, giving your answers as mixed numbers in their simplest form.

a $\frac{64}{12}$ **b** $\frac{124}{13}$

4 Write the numbers 2, 3 and 4 in the boxes to give the largest possible total. Each number can only be used once.

$$\square \frac{\square}{4} + \frac{1}{\square}$$

5 A rectangle has a length of 2.23 cm and a width of 4.55 cm, both to the nearest 0.01 cm.

a Work out the lower and upper bounds for the area of the rectangle in cm² to 3 decimal places.

b Give the area of the rectangle to a suitable degree of accuracy.

6 In 18 g of water there are 6.02×10^{23} molecules.

a How many molecules would there be in 10 g of water? Give your answer in standard form to 3 significant figures.

b Find the mass in kg of one molecule of water. Give your answer in standard form to 3 significant figures.

7 Work out an estimate for $(0.45 \times 0.78)^2$.

8 Find the value of (as an integer or fraction)

a 7^0 **b** $9^{\frac{1}{2}}$ **c** 8^{-2} **d** $64^{-\frac{1}{3}}$

9 Dylan rounds a number y to one decimal place. On rounding the number, the result is 5.6. Write down the error interval for y.

10 Show that the expression $\frac{1 - \sqrt{2}}{1 + \sqrt{2}}$ simplifies to $2\sqrt{2} - 3$.

11 The width of a sheet of glass, a, is 112 cm correct to 3 significant figures. Write down the error interval a.

12 Write the recurring decimal $0.\dot{7}\dot{2}$ as a fraction in its simplest form.

13 $c = \frac{a}{b}$

$a = 0.6754$ correct to 4 significant figures

$b = 2.34$ correct to 3 significant figures

a Work out the upper and lower bounds for a and b and hence work out the error interval for c. Give a reason for your answer.

b Give a value for c to a suitable degree of accuracy.

14 Work out each of the following. Simplify your answer where possible.

a $\frac{3}{25} \div \frac{9}{50}$ **b** $25 \div \frac{5}{16}$

15 Write each of the following numbers in standard form:

a 0.00000045 **c** 5640

b 12 million

16 Without using a calculator, work out the value of $(8 \times 10^{-5}) \times (4 \times 10^3)$ giving your answer in standard form.

17 a Write down all the factors of 64.

b Find the highest common factor of 64 and 100.

Algebra

Simple algebraic techniques

There are conventions (rules that everyone follows) when using algebra.

SNAP IT! Algebraic conventions

Write:
- ab instead of $a \times b$
- $3a$ instead of $a + a + a$ and $3 \times a$
- b^2 instead of $b \times b$
- a^3 instead of $a \times a \times a$
- a^2b instead of $a \times a \times b$
- $\frac{a}{b}$ instead of $a \div b$
- $\frac{1}{2}x$ or $\frac{x}{2}$ instead of $x \div 2$

Words used in maths

When using algebra certain words are used. Here are the ones you need to know about.

SNAP IT! Algebra terms

Equations contain letters and/or numbers and they always contain an equals sign. Examples include $2x + 4 = 2$, $3x - 4 = x + 6$, $x^2 + 2x + 1 = 0$.

Expressions contain letters and/or numbers but there is no equals sign. Examples include $2x - y^2$, $(x + 3)(x - 2)$.

Identities are true for all values of the letter or letters.

Examples include $(x + 1)^2 \equiv x^2 + 2x + 1$ or $(x - 2)^2 \equiv x^2 - 4x + 4$.

Formulae contain letters and/or words to represent variable quantities (**variables**) and they also contain an equals sign.

Terms are a single number or variable, or numbers and variables multiplied together. Terms are always separated by + or − signs. In the expression $5ab - 2a + 4$, $5ab$, $-2a$ and 4 are all terms.

Factors are what you can multiply together to get an expression in algebra. For example, $2x(x + 3) = 2x^2 + 6x$ so $2x$ and $(x + 3)$ are both factors of $2x^2 + 6x$.

> The symbol ≡ means 'is equivalent to'.
>
> Notice that the left-hand side is identical to the right-hand side, so this could not be solved like an equation.

Collecting like terms

Like terms are terms that have identical letters and powers. To simplify an expression you can collect like terms (find the sum of them). For example, $4x^2y$ can be combined with $-14x^2y$, but it cannot be combined with $-2xy^2$ or $5xy$.

WORKIT!

Simplify these expressions.

a $4x^2 + 6x - x + 6x^2 + 3$

$4x^2 + 6x - x + 6x^2 + 3 = 4x^2 + 6x^2 + 6x - x + 3$

$= 10x^2 + 5x + 3$

> Rearrange with the like terms grouped together.

> Always give answers in descending powers (e.g. x^3, then x^2, then x and finally ordinary numbers (e.g. 2)).

b $6a - b - c + a - 4a + c$

$6a - b - c + a - 4a + c = 6a + a - 4a - b - c + c = 3a - b$

> Always look at both the sign and the value of a letter or number

c $4yx + x^2 + 6xy - 4x^2$

$4yx + x^2 + 6xy - 4x^2 = 4xy + 6xy + x^2 - 4x^2 = 10xy - 3x^2$

> Although the order of the letters in a term does not matter ($4yx$ is the same as $4xy$), the convention is to write the number and then the letters in alphabetical order.

Substituting numerical values into formulae and expressions

This is the sort of calculation you often do in other subjects such as physics and chemistry. Replace each letter in the formula or expression with the numerical value given in the question and calculate.

WORKIT!

In the formula $p = \frac{nRT}{v}$, find the value of p if $n = 5$, $R = 8.31$, $v = 3$ and $T = 298$. Give your answer to 3 significant figures.

$p = \frac{nRT}{v} = \frac{5 \times 8.31 \times 298}{3}$

$= 4127.3$

$= 4130$ (to 3 s.f.)

> Write the equation and replace each letter by its value.

> Always check the question to see how many decimal places or significant figures should be used.

DOIT!

Write an example of an equation, expression, etc. on different cards. Write the terms on another set of cards. Practise matching the cards.

WORKIT!

Find the value of $b^2 - 4ac$ when $a = -1$, $b = -5$ and $c = 4$.

$b^2 - 4ac = (-5)^2 - 4(-1)(4)$

$= 25 + 16 = 41$

> Notice the use of BIDMAS.

NAILIT!

It is a good idea to add brackets around minus values in order to emphasise them.

✓ CHECKIT!

1 Identify whether each of the following is a formula, expression, equation or identity.

a $s = ut + \frac{1}{2}at^2$

b $2(x^2 + y^2) = 2x^2 + 2y^2$

c $4x^3y^2$

d $(x^2y)^3 = x^6y^3$

e $2x + 1 = 3$

2 Simplify

a $15x^2 - 4x + x^2 + 9x - x - 6x^2$

b $7a + 5b - b - 4a - 5b$

c $8yx + 5x^2 + 2xy - 8x^2$

d $x^3 + 3x - 5 + 2x^3 - 4x$

3 If $P = I^2R$, find P when $I = \frac{2}{3}$ and $R = 36$.

4 Using $v = u + at$, find v when $u = 20$, $a = -8$ and $t = 2$.

Removing brackets

Multiplying out a single bracket

Multiply each term inside the bracket by the term outside the brackets. Be careful when there is a negative outside the bracket as it will change the signs of all the terms inside the bracket.

WORKIT!

Multiply out the brackets and simplify your answer.

a $3(2x + 4)$

Multiply each term in the bracket by the number outside the bracket.

$3 \times 2x + 3 \times 4 = 6x + 12$

b $-6(x - 1)$

$-6 \times x + (-6) \times (-1) = -6x + 6$

The minus sign changes the sign of every term inside the bracket.

c $5(2x - 3) + 4(x - 1)$

$5 \times 2x + 5 \times (-3) + 4 \times x + 4 \times (-1)$
$= 10x - 15 + 4x - 4$
$= 14x - 19$

Multiply out the brackets, then collect like terms.

d $-(4 - x)$

$-(4 - x)$ is the same as $-1 \times (4 - x)$.

$-4 + x = x - 4$

Using the laws of indices with brackets

When multiplying out brackets you may have to use the laws of indices. A useful law is $x^a \times x^b = x^{a+b}$ so for example $x^3 \times x^2 = x^{3+2} = x^5$.

WORKIT!

Expand and simplify

a $2x(3x + 7)$

$2x \times 3x + 2x \times 7 = 6x^2 + 14x$

Remember that $x = x^1$.

b $4a(2a^2 - 3a + 4)$

$4a \times 2a^2 + 4a \times (-3a) + 4a \times 4 = 8a^3 - 12a^2 + 16a$

c $5x^2y^3z(3x + 7yz)$

$5x^2y^3z \times 3x + 5x^2y^3z \times 7yz = 15x^3y^3z + 35x^2y^4z^2$

Multiplying brackets

When multiplying pairs of brackets, all the terms in the first bracket must be multiplied by all the terms in the second bracket. There are different methods you can use to multiply two pairs of brackets together.

Method 1: the 'face' method

Using the 'face' method, the curved lines show that all the terms are multiplied together.

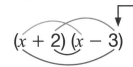

$$(x + 2)\ (x - 3)$$

Start from the top first and work from left to right:

- multiply x by x to give x^2
- multiply $+2$ by -3 to give -6
- multiply x by -3 to give $-3x$ ◄

  Now go to the bottom curved lines.
- multiply $+2$ by x to give $+2x$.

So, $(x + 2)(x - 3) = x^2 - 6 - 3x + 2x$
$$= x^2 - x - 6. ◄ $$

Notice that the answer is given in descending powers of x.

Method 2

Each term in the first bracket multiplies the contents of the second bracket:

$$(x + 2)(x - 3) = x(x - 3) + 2(x - 3)$$
$$= x^2 - 3x + 2x - 6 ◄ $$

Multiply out each bracket.

$$= x^2 - x - 6 ◄ $$

Simplify by collecting terms in x.

We will use Method 2 in this book, but you should use the method you are happiest with.

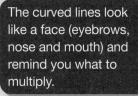

The curved lines look like a face (eyebrows, nose and mouth) and remind you what to multiply.

NAIL IT!

Try both methods and use the one you find easier.

NAIL IT!

With practice, you may find that you do not need to write the first step, but can go straight to multiplying out each bracket.

WORKIT!

Expand ◄ 'Expand' means 'multiply out'.

a $(x + 5)(x - 2)$

$x(x - 2) + 5(x - 2) = x^2 - 2x + 5x - 10$
$$= x^2 + 3x - 10 ◄ $$

Collect like terms in x.

b $(2x - 7)(4x - 2)$

$2x(4x - 2) - 7(4x - 2) = 8x^2 - 4x - 28x + 14 ◄ $
$$= 8x^2 - 32x + 14$$

Remember to keep the sign with the second term from the first bracket. You need to multiply each term in the second bracket by -7.

c $(a + b)^2$

$a(a + b) + b(a + b) = a^2 + ab + ab + b^2$
$$= a^2 + 2ab + b^2$$

$(a + b)^2$ means $(a + b)(a + b)$.

To expand an expression containing three sets of brackets, expand the first pair of brackets. Then expand this expression with the third bracket.

WORKIT!

Expand $(2x - 1)(x + 5)(3x - 2)$.

It is acceptable to use 1s in your working out.

$[2x(x + 5) - 1(x + 5)](3x - 2)$

$= (2x^2 + 10x - x - 5)(3x - 2)$ ← First expand $(2x - 1)(x + 5)$.

$= (2x^2 + 9x - 5)(3x - 2)$

$= 2x^2(3x - 2) + 9x(3x - 2) - 5(3x - 2)$ ← Multiply each term in $2x^2 + 9x - 5$ by each term in $3x - 2$.

$= 6x^3 - 4x^2 + 27x^2 - 18x - 15x + 10$

$= 6x^3 + 23x^2 - 33x + 10$

DOIT!

Write out the expansion of two brackets and make notes or write instructions beside each line of your working.

CHECKIT!

1 Multiply out

a $2(x + 4)$

b $7(9x + 3)$

c $-(1 - x)$

d $x(3x - 1)$

e $3x(x + 1)$

f $4x(5x - 2)$

2 Multiply out and simplify

a $2(x + 3) + 3(x + 2)$

b $6(x + 4) - 3(x - 7)$

c $x(3x + 1) + x(x + 1)$

d $x(3x - 4) - 2(3x - 4)$

3 Expand and simplify

a $(t + 3)(t + 5)$

b $(x - 3)(x + 3)$

c $(2y + 9)(3y + 7)$

d $(2x - 1)^2$

4 Expand and simplify

a $(x + 7)(x + 2)(2x + 3)$

b $(2x - 1)(3x - 2)(4x - 3)$

Factorising

Factorisation is the opposite process to expanding brackets.

Factorising a simple expression

Any factors that are common to all terms in an expression can be taken outside a bracket. These may be letters or numbers. When the terms include numbers (**coefficients**) the highest common factor of these is taken outside.

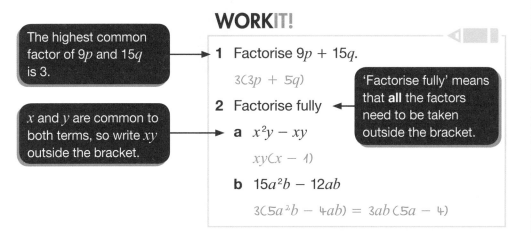

WORKIT!

The highest common factor of $9p$ and $15q$ is 3.

1 Factorise $9p + 15q$.

$3(3p + 5q)$

'Factorise fully' means that **all** the factors need to be taken outside the bracket.

2 Factorise fully

x and y are common to both terms, so write xy outside the bracket.

a $x^2y - xy$

$xy(x - 1)$

b $15a^2b - 12ab$

$3(5a^2b - 4ab) = 3ab(5a - 4)$

Factorising a quadratic expression

Factorising a quadratic expression gives two linear expressions in brackets. For example, $x^2 - 3x + 2$ factorises to $(x - 1)(x - 2)$.

Linear expressions contain the variable to the power 1 and no higher: for example, $2x - 7$.

There are many different methods for factorising a quadratic expression, one of which is shown below. This method is particularly useful when the numbers in the quadratic expression are large.

Suppose you are asked to factorise $x^2 + 7x - 30$.

You need to find two factors that multiply together to give -30. They also need to add together to give the coefficient of x ($+7$ in this case).

The coefficient of x is the number in front of x.

State the factor pairs of -30:

1 and -30	3 and -10	6 and -5
-1 and 30	-3 and 10	-6 and 5

Then look for the pair of numbers in the above list that add together to give 7:

-3 and 10.

With practice you may not need to write the list out.

Now write the two factors you have found as the coefficients of the x terms like this:

$x^2 + 7x - 30 = x^2 - 3x + 10x - 30$

Then factorise the first two terms by taking a term in x out and also the last two terms by taking a number out like this:

$$x^2 - 3x + 10x - 30 = x(x - 3) + 10(x - 3)$$

The term $(x - 3)$ is common, so may be taken out as a factor in the following way:

$$x(x - 3) + 10(x - 3) = (x - 3)(x + 10)$$

This may seem long-winded, but it works with all quadratic expressions that can be factorised, as the following examples show.

> **The terms inside the brackets have to be identical.**

> **If you had written the factors the other way round it would still work:**
> $x^2 + 10x - 3x - 30$
> $= x(x + 10) - 3(x + 10)$
> $= (x + 10)(x - 3)$.

WORKIT!

Factorise $x^2 - 9x + 18$.

$$x^2 - 9x + 18 = x^2 - 6x - 3x + 18$$
$$= x(x - 6) - 3(x - 6)$$
$$= (x - 6)(x - 3)$$

> The factors of 18 that add together to give -9 are -6 and -3. These are written as the coefficients of both x-terms.

> **The coefficient of x^2 is the number in front of x^2.**

If the coefficient of x^2 is not 1, multiply it by the number term. Then look for factor pairs of this number.

WORKIT!

Factorise $6x^2 - 17x + 12$.

$$6x^2 - 17x + 12 = 6x^2 - 9x - 8x + 12$$
$$= 3x(2x - 3) - 4(2x - 3)$$
$$= (2x - 3)(3x - 4)$$

> $6 \times 12 = 72$
> Look for a factor pair of 72 that adds to give -17: -9 and -8.

> Notice that the sign for the number outside the bracket is negative. This makes the contents of both brackets identical.

The difference of two squares

A perfect square is a number or expression whose square root does not contain a square root symbol. For example, 16, $36a^2$ and $100x^2y^4$ are all perfect squares; 10, $16b^3$ and $9xy^2$ are not perfect squares.

If two terms are both perfect squares with a minus sign between them, such as $x^2 - y^2$, this is called the difference of two squares. They can be factorised like this: $x^2 - y^2 = (x + y)(x - y)$.

Both terms must be perfect squares, so capable of being easily square rooted. Note that this only works when there is a minus sign between the two terms. For example:

$$4x^2 - 9y^2 = (2x + 3y)(2x - 3y)$$

$$16x^2 - 25 = (4x + 5)(4x - 5)$$

> ## NAIL IT!
>
> Always check your factorisation by multiplying out the brackets. It is easy to make a mistake, especially with signs.

> **Square root each term and put them in brackets: one bracket has a + sign between the terms and the other has a − sign.**

> ## SNAP IT!
>
> **Difference of two squares**
>
> Learn the formula for the difference of two squares:
>
> $$a^2 - b^2 \equiv (a + b)(a - b)$$

Factorising to simplify

Sometimes the question does not state that you need to factorise, but it is necessary in order to simplify an expression.

WORKIT!

Show that $\frac{1}{9x^2-1} \div \frac{1}{3x^2+20x-7}$ simplifies to $\frac{ax+b}{cx+d}$ where a, b, c and d are integers.

$$\frac{1}{9x^2-1} \div \frac{1}{3x^2+20x-7} = \frac{3x^2+20x-7}{9x^2-1}$$

Express the division as a single fraction.

$$= \frac{(3x-1)(x+7)}{(3x+1)(3x-1)}$$

Factorise both the numerator and the denominator.

$$= \frac{x+7}{3x+1}$$

Cancel the common factor.

DOIT!

Write a quick summary of how to factorise algebraic expressions.

CHECKIT!

1 Factorise fully

   a $24t + 18$

   b $9a - 2ab$

   c $5xy + 15yz$

   d $24x^3y^2 + 6xy^2$

2 Factorise

   a $x^2 + 10x + 21$

   b $x^2 + 2x - 15$

   c $6x^2 + 19x + 10$

   d $4x^2 - 49$

3 Work out

$$\frac{1}{x-7} - \frac{x+10}{2x^2-11x-21}$$

Express your answer in its simplest form.

Changing the subject of a formula

Changing the subject of a formula helps you to find a particular quantity. You need to get the letter representing that quantity on the left of the equals sign and everything else on the right. For example:

$$n = \frac{m}{M}$$

As n is on its own on the left-hand side, it is the **subject** of the formula. To change the subject of the formula to m, you need to remove M from the right-hand side. Multiply both sides by M to give:

$$Mn = \frac{Mm}{M}$$

Cancel the M in the top and bottom of the fraction on the right-hand side:

$$Mn = m$$

Swap this around to give:

$$m = Mn$$

To make M the subject of $n = \frac{m}{M}$, first multiply both sides by M to give:

$$Mn = m$$

Then remove the n from the left-hand side by dividing both sides by n.

So, $M = \frac{m}{n}$

> The upper case and lower case letters mean different things, so they are separate terms.

> We normally write the letters in an equation in alphabetical order: Mn rather than nM.

> m is now the subject.

> M is now the subject.

WORKIT!

Rearrange the following equations to make the bracketed letter the subject of the formula.

a $y = \frac{1}{x}$ (x)

$$xy = 1$$

$$x = \frac{1}{y}$$

b $y = 2x + 6$ (x)

$$y - 6 = 2x$$

$$x = \frac{y - 6}{2}$$

c $r^2 = a^2 + b^2$ (b)

$$r^2 - a^2 = b^2$$

$$b^2 = r^2 - a^2$$

$$b = \sqrt{r^2 - a^2}$$

> Multiply both sides by x to remove x from the denominator.

> Divide both sides by y.

> Subtract 6 from both sides.

> Divide both sides by 2.

> Subtract a^2 from both sides.

> Square root both sides.

NAILIT!

Always put the subject of the formula on the left-hand side.

Rearranging a formula for a term that appears on both sides

SNAP IT!   Rearranging formulae

When the new subject appears on both sides of a formula:

① If there are any brackets, multiply them out.

② Get all the terms that contain the subject on one side of the equation and all the other terms on the other side.

③ Take the new subject out as a factor so you will have it multiplied by the contents of a bracket.

④ Divide both sides by the bracket, leaving the subject on its own on the left of the equation.

WORKIT!

Make y the subject of the formula $3(3x - y) = 9 - 4xy$.

Collect the terms containing the new subject on one side and all the other terms on the other side.

$$9x - 3y = 9 - 4xy$$

Multiply out the brackets.

$$4xy - 3y = 9 - 9x$$

$$y(4x - 3) = 9 - 9x$$

Take the new subject, y, out of a bracket as a factor.

Divide both sides by the contents of the bracket to leave the new subject on its own.

$$y = \frac{9 - 9x}{4x - 3}$$

DOIT!

Draw a flowchart for rearranging a formula when the subject appears on both sides.

CHECKIT!

1 Make r the subject of the formula.

  a $A = \pi r^2$

  b $A = 4\pi r^2$

  c $V = \frac{4}{3}\pi r^3$

2 Make the bracketed symbol the subject of the formula.

  a $y = mx + c$ (c) **d** $v^2 = 2as$ (s)

  b $v = u + at$ (u) **e** $v^2 = u^2 + 2as$ (u)

  c $v = u + at$ (a) **f** $s = \frac{1}{2}(u + v)t$ (t)

Solving linear equations

Linear equations are ones where the unknown or **variable** doesn't have a power (except 1). The variable is usually called x, but it can be any letter or symbol. You only need one linear equation to solve one variable.

Remember that an equation is like a balance. If you apply a process to one side (such as add a number), then for the equation to remain in balance, the same process (adding the same number) must be applied to the other side.

WORKIT!

Solve these equations.

You want to get x on its own on the left-hand side.
Subtract 5 from both sides.

To remove the 4 in front of the x, divide both sides by 4.

Remove the denominator by multiplying both sides by 3.

First, get a term in x on its own.
Add 7 to both sides.

a $x + 5 = 7$

$x + 5 - 5 = 7 - 5$

$x = 2$

b $4x = 24$

$x = 24 \div 4 = 6$

c $4x - 7 = 21$

$4x = 28$

$x = 7$

d $\frac{2x - 7}{3} = 7$

$2x - 7 = 21$

$2x = 28$

$x = 14$

Linear equations involving brackets

When solving linear equations containing brackets, first multiply out the brackets and collect any like terms. Then solve the equation in the way outlined in the previous examples.

WORKIT!

Solve these equations.

Multiply out the brackets.

Multiply out the brackets and collect like terms.

a $4(2x - 3) = 4$

$8x - 12 = 4$

$8x = 16$

$x = 2$

b $5(x - 3) - 3(x + 1) = 0$

$5x - 15 - 3x - 3 = 0$

$2x - 18 = 0$

$2x = 18$

$x = 9$

NAILIT!

It is better to rearrange the equation in such a way that the unknown quantity is positive.

Unknown quantity on both sides of the equation

If the unknown quantity (i.e. the quantity you are asked to find) appears on both sides of the equation, you must get it on just one side.

WORKIT!

Solve these equations.

a $2x - 1 = x + 4$

$$x - 1 = 4$$

$$x = 5$$

You could subtract $2x$ from both sides but this would give $-x$ on the right-hand side. Subtracting x from both sides is a better option.

b $3(3x - 5) = 12(x - 7)$

$$9x - 15 = 12x - 84$$

Subtract $9x$ from both sides.

$$-15 = 3x - 84$$

$$69 = 3x$$

Subtract 84 from both sides.

$$23 = x$$

$$x = 23$$

NAILIT!

Solving linear equations is very important in GCSE mathematics as it happens in different kinds of questions. It is essential that you master all the techniques here as they will be needed elsewhere.

DOIT!

Make up some linear equations.
Solve them tomorrow. Are your answers all integers?

 # CHECKIT!

1 Solve these equations.

a $x - 7 = -4$ $x = 3$ $x = -4 + 7$

c $\frac{x}{5} = 4$ $x = 4 \times 5$ $x = 20$

b $9x = 27$ $x = \frac{27}{9}$ $x = 3$

2 Solve these equations.

a $3x + 1 = 16$

c $\frac{3x}{5} + 4 = 16$

b $\frac{2x}{3} = 12$

3 Solve these equations.

a $5(1 - x) = 15$

b $2m - 4 = m - 3$

c $9(4x - 3) = 3(2x + 3)$

Solving quadratic equations using factorisation

A quadratic equation is an equation that can be expressed in the form:

$$ax^2 + bx + c = 0$$

where a, b and c are numbers.

To solve a quadratic using factorisation, first put the equation into the form above and then factorise it. For example, to solve $2x^2 = 5 - 9x$, first rearrange it as $2x^2 + 9x - 5 = 0$ and then factorise it to give:

Before factorising the quadratic expression must be equal to zero.

$$(2x - 1)(x + 5) = 0$$

For this to equal zero, one of the brackets must equal zero:

Hence $2x - 1 = 0$ or $x + 5 = 0$,

giving $x = \dfrac{1}{2}$ or $x = -5$

NAILIT!

You need to be able to solve simple linear equations to solve quadratic equations by factorisation.

WORKIT!

Solve $2x^2 + 7x - 4 = 0$

$(2x - 1)(x + 4) = 0$ ⟵ Factorise $2x^2 + 7x - 4$.

$2x - 1 = 0$ or $x + 4 = 0$ ⟵ Put each bracket equal to 0.

$x = \dfrac{1}{2}$ or $x = -4$

WORKIT!

This rectangle has an area of 60. Find the value of x.

Area of rectangle = length × width

$$= 6x(2x + 1)$$

Substitute area = 60. ⟶ $60 = 6x(2x + 1)$

$$60 = 12x^2 + 6x$$

$$0 = 12x^2 + 6x - 60$$

$12x^2 + 6x - 60 = 0$ ⟵ Divide both sides by 6 to make the factorisation easier.

$$2x^2 + x - 10 = 0$$

$$(2x + 5)(x - 2) = 0$$

$x = -\dfrac{5}{2}$ or $x = 2$ ⟵ Put each bracket equal to 0 and solve.

$x = -\dfrac{5}{2}$ would give a length of $6 \times \left(-\dfrac{5}{2}\right) = -15$, which is impossible.

So $x = 2$

(Rectangle labelled: top side $6x$, right side $2x + 1$)

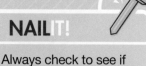

NAILIT!

Always check to see if both values are possible or only one of them. In some situations, you can ignore a negative value, for example if it is a length.

WORKIT!

The rectangle shown below has length $(x + 8)$ cm and width $(x - 2)$ cm.
The area of the rectangle is 56 cm^2. Find the value of x.

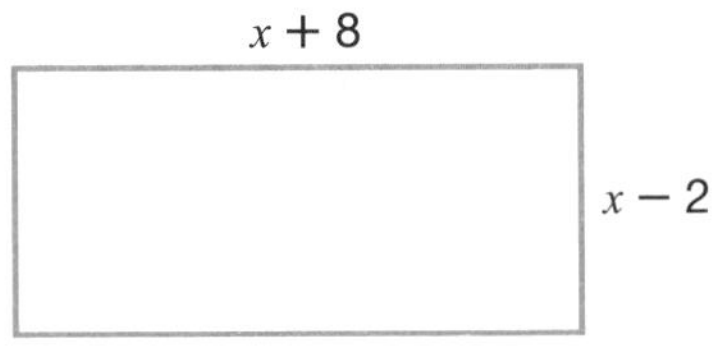

Area $= (x + 8)(x - 2) = x^2 + 6x - 16$

Also, area $= 56$ so $x^2 + 6x - 16 = 56$

Hence $x^2 + 6x - 72 = 0$

Factorising we obtain $(x - 6)(x + 12) = 0$ so $x = 6$ or -12

$x = -12$ would give a negative length and width so this is impossible.

So $x = 6$ cm

$x + 8$

$x - 2$

DO IT!

Draw a poster showing the solution of a quadratic equation using factorisation, annotating the stages.

✓ CHECKIT!

1 Solve

$(x + 5)(x + 1)$

a $x^2 + 5x + 6 = 0$

b $x^2 - x - 12 = 0$

c $2x^2 + 17x + 35 = 0$

2 A right-angled triangle has base $(2x + 3)$ cm and height $(x + 4)$ cm. The area of the triangle is 9 cm^2.

a Use the area of the triangle to show that $2x^2 - 11x - 6 = 0$.

b Solve this equation to find the only possible value of x.

c Work out the base and height of the triangle.

3 Three sides of a right-angled triangle are 13 cm, $(x + 1)$ cm and $(x + 8)$ cm. The longest side is the 13 cm side.

Calculate the value of x.

Solving quadratic equations using the formula

NAILIT!

This formula will not be given so you will need to remember it.

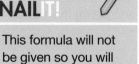

Not all quadratic equations can be factorised easily. There are three other methods you can use: using a formula, completing the square or drawing a graph. Completing the square is covered on page 79 and finding solutions graphically on pages 80–2.

NAILIT!

Check your answers by substituting the values back into the equation.

SNAPIT!   Quadratic formula

Quadratic equations of the form $ax^2 + bx + c = 0$ can be solved using the formula:

$$x = \frac{-b \pm \sqrt{b^2 - 4ac}}{2a}$$

Be careful with signs when you are entering numbers into this formula.

NAILIT!

If you are asked to leave your answer in surd form, then you must use either the formula or the method involving completing the square (see page 79).

WORKIT!

Solve the equation $2x^2 - 5x + 1 = 0$ leaving your answers in surd form.

$x = \dfrac{-b \pm \sqrt{b^2 - 4ac}}{2a}$ ← Write down the formula.

Comparing the equation with $ax^2 + bx + c = 0$ gives $a = 2$, $b = -5$ and $c = 1$.
Substitute these values into the formula.

$x = \dfrac{-(-5) \pm \sqrt{(-5)^2 - 4(2)(1)}}{2(2)}$

$= \dfrac{5 \pm \sqrt{25 - 8}}{4} = \dfrac{5 \pm \sqrt{17}}{4}$

So $x = \dfrac{5 + \sqrt{17}}{4}$ or $\dfrac{5 - \sqrt{17}}{4}$

WORKIT!

a Show that $\dfrac{3}{x + 7} = \dfrac{2 - x}{x + 1}$ can be written as $x^2 + 8x - 11 = 0$.

$\dfrac{3}{x + 7} = \dfrac{2 - x}{x + 1}$ ← Multiply both sides by $(x + 1)(x + 7)$.

$3(x + 1) = (2 - x)(x + 7)$

$3x + 3 = 2x + 14 - x^2 - 7x$ ← Expand the brackets.

$3x + 3 = -x^2 - 5x + 14$

$x^2 + 8x - 11 = 0$ ← Rearrange in the form $ax^2 + bx + c = 0$.

b Hence solve the equation
$\frac{3}{x + 7} = \frac{2 - x}{x + 1}$, giving your answers to 2 decimal places.

$x = \dfrac{-b \pm \sqrt{b^2 - 4ac}}{2a}$

$= \dfrac{-8 \pm \sqrt{8^2 - 4(1)(-11)}}{2(1)}$ ⟵ Substitute $a = 1$, $b = 8$ and $c = -11$ into the formula.

$= \dfrac{-8 \pm \sqrt{64 + 44}}{2}$

$= \dfrac{-8 \pm \sqrt{108}}{2}$

$= \dfrac{-8 + \sqrt{108}}{2}$ or $\dfrac{-8 - \sqrt{108}}{2}$

$= 1.1962$ or -9.1962

$= 1.20$ or -9.20 (to 2 d.p.)

NAILIT!

If the question asks for the answer to be given to a certain number of decimal places, you need to use the formula rather than factorisation.

DOIT!

Use the quadratic formula to solve the equations you solved by factorisation in the previous section.

CHECKIT!

$-b \pm \sqrt{b^2 + 4ac} \div 2a$

1 Solve the equation $2x^2 - x - 7 = 0$, giving your answers correct to 3 significant figures.

$a = 2 \quad b = -1 \quad c = -7$

$\dfrac{1 \pm \sqrt{-1^2 + 4(2)(7)}}{2(2)}$

$\dfrac{1 \pm \sqrt{1 + 56}}{4} = \dfrac{1 \pm \sqrt{57}}{4}$

$x = 2.8 \quad x = -0.8$

2 a Show that $\frac{2x + 3}{x + 2} = 3x + 1$ can be written as $3x^2 + 5x - 1 = 0$.

b Hence solve the equation $\frac{2x + 3}{x + 2} = 3x + 1$, giving your answers to 2 decimal places.

Solving simultaneous equations

Simultaneous equations are sets of equations that have more than one variable (e.g. x and y). Solving the equations means finding values for x and y that are true for both equations.

SNAPIT!   Simultaneous equations

There are three methods that can be used to solve a pair of simultaneous equations:

1. **By elimination**: eliminate one of the unknowns by adding or subtracting the two simultaneous equations.

2. **By substitution**: substitute the expression for x or y from one of the equations into the other equation.

3. **Graphically**: use both equations to plot two lines – their point of intersection is the solution.

Solving simultaneous equations by elimination

Eliminate one of the variables by either adding or subtracting the equations.

WORKIT!

Solve the simultaneous equations.

$$2x + 3y = 8 \qquad (1)$$
$$x + 4y = 9 \qquad (2)$$
$$2x + 8y = 18 \qquad (3)$$
$$5y = 10$$
$$y = 2$$
$$2x + 3y = 8 \qquad (1)$$
$$2x + 3(2) = 8$$
$$2x + 6 = 8$$
$$2x = 2$$
$$x = 1$$
$$x + 4y = 9 \qquad (2)$$
$$1 + 4(2) = 9$$
Solution is $x = 1$ and $y = 2$.

Number the equations so it is easier to refer to them.

Subtract (1) from (3).

Make the coefficients of either x or y the same for both equations, by multiplying one or both of the equations by numbers.
In this case, multiply equation (2) by 2 to make the coefficients of x the same.

Substitute $y = 2$ into equation (1).

Check the answer by substituting $x = 1$ and $y = 2$ into equation (2).

NAILIT!

Always check your answer by substituting into the equations.

Solving simultaneous equations by substitution

This method of solving simultaneous equations involves removing y by substituting the y value from one of the equations into the other equation. The substitution method is preferred when finding the solutions to a ← non-linear equation and a linear equation.

> Non linear means it has a power of x, as in a quadratic equation.

WORKIT!

Solve the simultaneous equations.

$x^2 + y^2 = 16$ ←

> The first equation represents a circle because it is the form $x^2 + y^2 = r^2$. The second equation is a straight line because it is in the form $ax + by = c$.

$x + y = 4$

$x = 4 - y$ ←

> Rearrange the second equation to express x in terms of y.

$(4 - y)^2 + y^2 = 16$ ←

$16 - 8y + y^2 + y^2 = 16$

> Now substitute for x into the first equation.

$2y^2 - 8y = 0$ ←

> Subtract 16 from both sides.

$2y(y - 4) = 0$ ←

> Remove $2y$ as a factor.

$2y = 0$ giving $y = 0$, or $y - 4 = 0$ giving $y = 4$.

Substituting $y = 0$ into $x + y = 4$ gives $x = 4$. ←

> You can use either equation but this one is easier.

Substituting $y = 4$ into $x + y = 4$ gives $x = 0$.

So the solutions are $x = 4$ and $y = 0$, or $x = 0$ and $y = 4$.

NAILIT!

Make sure you write 'or' between the two sets of values, not 'and'.

Solving simultaneous equations graphically

Any pair of equations can be solved graphically by drawing them accurately and finding any point or points of intersection.

NAILIT!

When asked to plot a graph, create a table of values.

WORKIT!

Solve these simultaneous equations graphically, giving your solutions to 1 decimal place.

$y = x^2 - 4x + 1$

$y = x - 2$

x	0	1	2	3	4	5
$y = x^2 - 4x + 1$	1	-2	-3	-2	1	6

> This is a quadratic equation, so you need five or six pairs of coordinates to plot the graph.

x	0	2	5
$y = x - 2$	-2	0	3

> You only need two pairs of coordinates for a straight line, but a third one provides a check.

Plot the graphs of the two equations and find where they intersect.

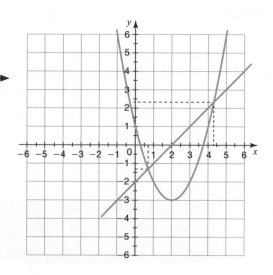

NAILIT!

Two linear equations could be solved in a similar way, but there would be only one point of intersection so only one pair of values.

The two pairs of solutions are:

$x = 4.3, y = 2.3$ and $x = 0.7, y = -1.3$

DOIT!

Create revision cards to show when each method of solving simultaneous equations is most appropriate.

STRETCHIT!

Not all simultaneous equations can be solved. If lines are parallel to each other there are no points of intersection.

CHECKIT!

1 Solve these pairs of simultaneous equations.

    a $y = 3x - 7$ **b** $y = 2x - 6$

       $y = 3 - 2x$ $y = -3x + 14$

2 Solve the simultaneous equations.

   $y = 10x^2 - 5x - 2$

   $y = 2x - 3$

3 The line $y = 6x + 2$ and the curve $y = x^2 + 5x - 4$ intersect at two points. Use an algebraic method to find the coordinates of these points.

Solving inequalities

Representing inequalities

Number lines and set notation

Both number lines and set notation can be used to represent inequalities, as shown below.

In set notation, { } contains the possible set of values.

∪ means that the set of numbers contains all the numbers represented by both inequalities.

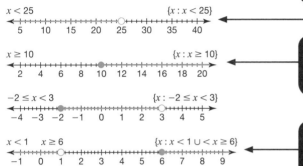

$x < 25$ $\{x : x < 25\}$

$x \geq 10$ $\{x : x \geq 10\}$

$-2 \leq x < 3$ $\{x : -2 \leq x < 3\}$

$x < 1$ $x \geq 6$ $\{x : x < 1 \cup < x \geq 6\}$

The blue line shows x is less than 25. The open circle at 25 shows that x cannot be equal to 25.

The blue line shows x is 10 or greater. The closed circle at 10 shows that x can be equal to 10.

When there is a gap between the circles, two separate inequalities are represented.

Graphically

Inequalities can be shown graphically using boundary lines and shading. If the inequality allows points on the boundary line, then it is drawn as a solid line; otherwise it is a dotted line.

WORKIT!

Shade the region of points on a graph where

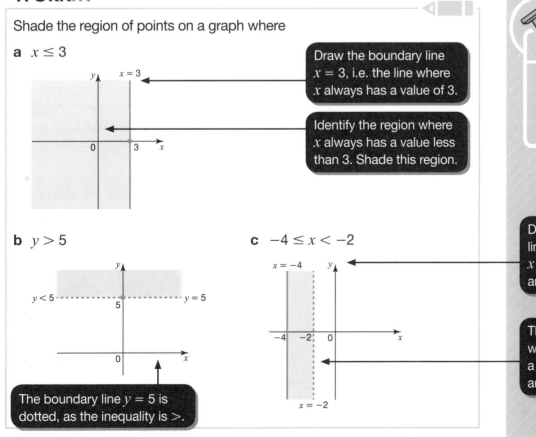

a $x \leq 3$

Draw the boundary line $x = 3$, i.e. the line where x always has a value of 3.

Identify the region where x always has a value less than 3. Shade this region.

b $y > 5$

The boundary line $y = 5$ is dotted, as the inequality is >.

c $-4 \leq x < -2$

Draw both boundary lines. Remember that $x = -4$ needs to be solid and $x = -2$ dotted.

The shaded region is where x always has a value between -4 and -2.

NAIL IT!

Always remember that the region that is allowed is the shaded region.

Linear inequalities

Solve inequalities in the same way as you solve equations, by doing the same operation to both sides of the inequality.

WORKIT!

Solve these inequalities.

a $4x - 5 > 2$

$4x > 7$

$x > \frac{7}{4}$

Add 5 to both sides.

Divide both sides by 4.

b $3(2x - 7) \geq 12x + 15$

$6x - 21 \geq 12x + 15$

$-6x - 21 \geq 15$

$-6x \geq 36$

$x \leq -6$

c $-3 \leq 2x + 1 < 7$

$-4 \leq 2x < 6$

$-2 \leq x < 3$

Subtract 1 from all three elements.

Divide all three elements by 2.

When you multiply or divide by a negative number, reverse the inequality.

Linear inequalities with two variables

Sometimes you will be given two or three equations with two variables and asked to show the solution on a graph:

To draw a straight line, substitute two values of x into the equation to find y, and draw the line through the two points.

1 Draw the lines representing boundary lines by replacing the inequality symbol with $=$. If the inequality is $\leq$ or $\geq$ make the line solid; if it is $<$ or $>$ make the line dotted.

If you are not sure which side of the line to shade, substitute a pair of values for x and y.

2 For each line, shade the region that is allowed.

WORKIT!

a Illustrate the region represented by the following inequalities on a graph by shading the region that is required.

$y \leq x$ $x + 3y \leq 12$ $y > 1$

b List the integer values of x and y that satisfy all of these inequalities.

Draw the lines $y = x$, $x + 3y = 12$ as solid lines and $y = 1$ as a dotted line.

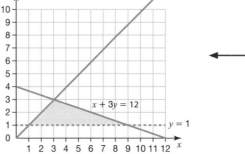

To work out which side of $x + 3y = 12$ to shade, substitute $x = 1$ and $y = 4$: $1 + 3(4) = 13$. The inequality is $x + 3y \leq 12$, so (1, 4) is not allowed.

Any point in the required region or on solid lines enclosing the region satisfies all three inequalities.

$(2, 2), (3, 2), (4, 2), (5, 2), (6, 2), (3, 3)$

Quadratic inequalities

To solve a quadratic inequality such as $x^2 + x \geq 6$:

1 Ensure that the quadratic inequality has all the terms on one side and a zero on the other side: $x^2 + x - 6 \geq 0$.

2 Consider the case where $x^2 + x - 6 = 0$.
Solve the equation:
$x^2 + x - 6 = (x + 3)(x - 2) = 0$, ◄——— These are the points where the curve cuts the x-axis.
so $x = -3$ or $x = 2$.

3 Sketch the curve $x^2 + x - 6$.

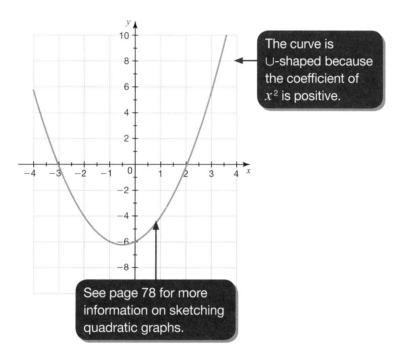

The curve is U-shaped because the coefficient of x^2 is positive.

See page 78 for more information on sketching quadratic graphs.

4 Identify the values of x where $x^2 + x - 6 \geq 0$.
These are the values of x for which y is greater than or equal to 0 – where the graph is above or on the x-axis: $x \leq -3$ and $x \geq 2$.

For the opposite inequality ($x^2 + x - 6 \leq 0$), the values of x would be for the graph on or below the x-axis: $-3 \leq x \leq 2$.

WORKIT!

a Solve the inequality $x^2 - 7x + 10 < 0$.

$x^2 - 7x + 10 = 0$ gives $(x - 2)(x - 5) = 0$, so the curve
intersects the x-axis at $x = 2$ and $x = 5$. ◄─── Work out the solutions
to $x^2 - 7x + 10 = 0$
and sketch the curve,
marking the roots of
the quadratic.

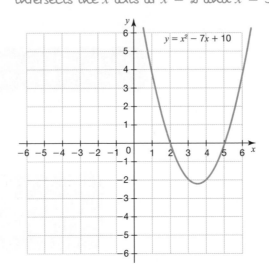

Solution is $2 < x < 5$. ◄─── Identify the values of x
for the graph below the
x-axis because of the
$<$ in the inequality.

b Show the solutions on a number line.

DOIT!

Make up some
inequalities, rolling
dice to generate the
numbers. Solve them.

CHECKIT!

1 Solve these inequalities. Show your
answers on a number line and using set
notation.

   a $1 - 2x < -11$ **c** $\dfrac{x - 5}{3} < 7$

   b $2x - 7 \geq 15$

2 Solve these inequalities.

   a $2x - 4 > x + 6$

   b $4 + x < 6 - 4x$

   c $2x + 9 \geq 5(x - 3)$

3 a Illustrate the region represented by the
following inequalities on a graph.

    $x > -1$ $y \leq 2$ $x - y < 2$

   b List the integer values of x and y that
satisfy all of these inequalities.

4 Solve the inequality $x^2 > 3x + 10$.

Problem solving using algebra

Problems involving simultaneous equations

In some questions you will have to form the two simultaneous equations from information given in the question and then solve them.

WORKIT!

The sum of two numbers is 14 and the difference of the same numbers is 2. Find the two numbers.

Let the larger number $= x$ and the smaller number $= y$.

$$x + y = 14 \qquad (1)$$

$$x - y = 2 \qquad (2)$$

> Create the simultaneous equations from the information in the question.

> Add equations (1) and (2).

$$2x = 16$$

$$x = 8$$

> Substitute $x = 8$ into equation (1).

$$x + y = 14$$

$$8 + y = 14$$

$$y = 6$$

Check:

$$x - y = 2 \qquad (2)$$

> Substitute $x = 8$ and $y = 6$ into equation (2).

$$8 - 6 = 2$$

$$2 = 2$$

> Both sides of the equation are equal, showing that the values of x and y satisfy equation (2).

The two numbers are 6 and 8.

Problems involving inequalities

In some problems, you will be given a maximum or minimum value (e.g. length, area, volume). You then set up an inequality and solve it.

WORKIT!

This solid cuboid has length $4x$ cm, width $2x$ cm and height x cm, where x is an integer.

The total surface area of the cuboid is less than or equal to 2800 cm².

Show that $x \leq 10$.

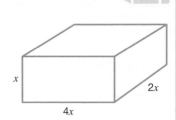

Area of top + bottom = $8x^2 + 8x^2 = 16x^2$

Area of front + back = $4x^2 + 4x^2 = 8x^2$

Area of the two sides = $2x^2 + 2x^2 = 4x^2$

Total surface area = $16x^2 + 8x^2 + 4x^2 = 28x^2$

> Find an expression for the total surface area in terms of x.

$28x^2 \leq 2800$

> Express the total surface area as an inequality.

$x^2 \leq 100$

$x \leq 10$

> $\sqrt{100} = \pm 10$, but it would not make sense for x to be negative when finding the sides of the cuboid.

DOIT!

Design a poster to show how to turn a word problem into an algebraic expression.

CHECKIT!

1 Find two numbers such that their sum is 77 and their difference is 25.

2 A number is added to the numerator and denominator of the fraction $\frac{15}{31}$.

The resulting fraction is $\frac{5}{6}$. Find the number.

3 The perimeter of a rectangle is 24 cm and its area is 27 cm².

Find the length and width of the rectangle.

Use of functions

A **function** takes a number as **input** and gives another number as **output**.

The function is represented as a box and it does something mathematical to the number you input. Whatever the input it performs the same mathematical operation or operations on it to give the output.

This function can be written as
$f(x) = \frac{x}{3} + 1$.

When applied to the number 18,
it is expressed as $f(18) = \frac{18}{3} + 1 = 7$.

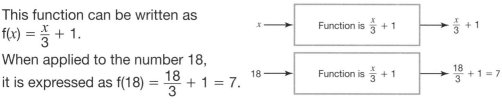

Similarly, the function 'square x and then add x' can be written as $f(x) = x^2 + x$.

When applied to the number 3, it is expressed as $f(3) = 3^2 + 3 = 12$. ◄——— Replace x with 3.

WORKIT!

$f(x) = x^2 - 2x + 1$

Work out

a f(2) ◄——————————— Substitute $x = 2$ in $x^2 - 2x + 1$.

$f(2) = 2^2 - 2(2) + 1 = 1$

b f(−1)

$f(-1) = (-1)^2 - 2(-1) + 1 = 4$ ◄——— Put brackets around negative numbers.

c the values of x for which $f(x) = 0$.

When $f(x) = 0$, $x^2 - 2x + 1 = 0$ ◄——— This is a quadratic equation that needs to be factorised and then solved.

$(x - 1)(x - 1) = 0$

So $x = 1$

Inverse functions

The **inverse** of a function will produce the input value from the output value. The inverse of the function $f(x)$ is written as $f^{-1}(x)$.

SNAPIT! Inverse functions

To find the inverse of a function:

1 Let the function equal y.

2 Rearrange the resulting equation so that x is the subject of the equation.

3 Replace x with $f^{-1}(x)$ and replace y with x.

WORKIT!

$f(x) = \frac{x}{3} + 1$. Find $f^{-1}(x)$.

Let $y = \frac{x}{3} + 1$

$3y = x + 3$ ◀ Rearrange to make x the subject. First multiply both sides by 3 to remove the denominator.

$x = 3y - 3$

Replace x with $f^{-1}(x)$ and replace y with x.

$f^{-1}(x) = 3x - 3$

Composite functions

Composite functions involve applying two or more functions in succession.

Consider the following example:

$f(x) = x^2$ and $g(x) = x - 2$

The composite function $fg(x)$ means $f(g(x))$ and is the result of performing the function g first and then the function f.

Here, g means 'subtract 2 from it' and f means 'square it'. So fg means 'subtract 2 from it' and then 'square it', i.e. $(x - 2)^2$. gf means 'square it' and then 'subtract 2 from it', i.e. $x^2 - 2$.

To find $fg(x)$, replace x by the expression for $g(x)$ in $f(x)$:

$fg(x) = f(g(x)) = f(x - 2) = (x - 2)^2$

WORKIT!

The functions f and g are such that $f(x) = 3(x + 1)$ and $g(x) = \frac{x}{3} + 2$.

Find

a g(12)

$g(12) = \frac{12}{3} + 2 = 4 + 2 = 6$

b $g^{-1}(x)$

Let $y = \frac{x}{3} + 2$

$3y = x + 6$

$x = 3y - 6$

$g^{-1}(x) = 3x - 6$

c fg(x)

$fg(x) = 3\left(\left(\frac{x}{3} + 2\right) + 1\right)$

$= 3\left(\frac{x}{3} + 3\right)$

$= x + 9$

DOIT!

Find an inverse function, using both the function machine method and the algebraic method.

CHECKIT!

1 $f(x) = \frac{1}{x-1}$ where $x \neq 1$.

Find

a f(0)   **b** $f\left(-\frac{1}{2}\right)$ **c** $f^{-1}(x)$

2 $f(x) = \sqrt{(x^2 - 9)}$ and $g(x) = x + 4$.

Find

a fg(x)   **b** gf(x)   **c** gf(3)

Iterative methods

Iteration means repeatedly performing a function, each time using your previous answer to get the next result. Iterative methods enable equations to be solved that would be hard to solve using other methods.

Using changes of sign of f(x) to locate solutions

If f(x) can take any value between a and b, and there is a change of sign between f(a) and f(b), then a solution of f(x) = 0 lies between a and b.

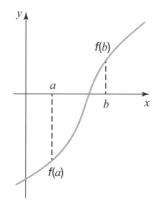

WORKIT!

Show that the equation

$9x^3 - 9x + 1 = 0$

has a solution between 0 and 0.2. ◄—— a and b are 0 and 0.2, respectively.

Let $f(x) = 9x^3 - 9x + 1$

$f(0) = 1$

$f(0.2) = 9(0.2)^3 - 9(0.2) + 1 = 0.072 - 1.8 + 1 = -0.728$

As there is a sign change, there is a solution between 0 and 0.2.

Insert each value for x into the function and see if there is a sign change between f(0) and f(0.2).

Entering values into a recurrence relation to solve an equation

A recurrence relation gives a value in terms of the previous value. Recurrence relations can be used to solve equations, i.e. find a **root**. ◄——

A root of an equation is a value of x where the function equals 0, i.e. where its graph crosses the x-axis.

WORKIT!

The subscript tags n and $n + 1$ mean that there is a series of values for x, and each solution is used to generate the next one, using this function. We start with $n = 0$.

The recurrence relation

$$x_{n+1} = \left(\frac{2x_n + 5}{4}\right)^{\frac{1}{3}}$$

with $x_0 = 1.2$, may be used to find a, a root of the equation $4x^3 - 2x - 5 = 0$.

a is one of the roots of the equations (i.e. a solution).

Find and record the values of x_1, x_2, x_3 and x_4.

Write down the value of x_4 correct to 5 decimal places and prove that this value is the value of the root a correct to 5 decimal places.

$$x_0 = 1.2$$

To get x_1, substitute the value for x_0 into the equation. Then substitute x_1 to find x_2, and so on.

$$x_1 = \left(\frac{2x_0 + 5}{4}\right)^{\frac{1}{3}} = \left(\frac{2(1.2) + 5}{4}\right)^{\frac{1}{3}} = 1.227601026$$

$$x_2 = \left(\frac{2x_1 + 5}{4}\right)^{\frac{1}{3}} = \left(\frac{2(1.227601026) + 5}{4}\right)^{\frac{1}{3}} = 1.230645994$$

$$x_3 = \left(\frac{2x_2 + 5}{4}\right)^{\frac{1}{3}} = \left(\frac{2(1.230645994) + 5}{4}\right)^{\frac{1}{3}} = 1.230980996$$

$$x_4 = \left(\frac{2x_3 + 5}{4}\right)^{\frac{1}{3}} = \left(\frac{2(1.230980996) + 5}{4}\right)^{\frac{1}{3}}$$

Do not round these numbers yet.

$$= 1.231017841$$

Round the final answer to the required number of decimal places.

$$x_4 = 1.23102 \text{ (to 5 d.p.)}$$

The root lies on the x-axis where $f(x) = 0$. Test x values either side of your answer by putting them into the function. If one output is positive and one negative, your answer is in the right range and is correct.

DOIT!

Draw a flowchart for using iteration to find a root to a specified accuracy.

$$f(1.231015) = 4(1.231015)^3 - 2(1.231015) - 5 = -0.000119668$$

$$f(1.231025) = 4(1.231025)^3 - 2(1.231025) - 5 = 0.000042182$$

As there is a sign change between these values of $f(x)$, 1.23102 is in the correct range.

CHECKIT!

1 The cubic equation $x^3 - x - 2 = 0$ has a root a between 1 and 2.

The recurrence relation

$$x_{n+1} = (x_n + 2)^{\frac{1}{3}}$$

with $x_0 = 1.5$ can be used to find a.

Calculate x_4, giving your answer correct to 3 decimal places. Prove that this value is also the value of a correct to 3 decimal places.

Equation of a straight line

Straight-line graphs are also called **linear** graphs and have an equation of the form:

$y = mx + c$ ← There is a single y on the left-hand side of the equation.

m is the **gradient** (i.e. the steepness of the line) and c is the **intercept on the y-axis** (where the graph cuts the y-axis).

It is important that in this equation there is only an x term. There are no terms containing x^2, x^3, $\sqrt{x}$, $\frac{1}{x}$, and so on.

This is an equation of a straight line: $y = 2x - 3$.

Comparing this equation with $y = mx + c$, the gradient, m, is 2 and the intercept on the y-axis, c, is -3.

NAILIT!

For an equation to be that of a straight line, you must be able to rearrange it in the form $y = mx + c$ (with no higher power than x).

WORK**IT!**

Which of these equations are straight lines? Explain your answer.

a $y = 1.4x + 7$

Yes: in the form $y = mx + c$.

b $y = 4x^2 + 1$

No: not in the form $y = mx + c$ as it contains a term in x^2.

c $y = \dfrac{2}{x}$

No: not in the form $y = mx + c$.

d $5x + 2y = 3$

Yes: can be rearranged to give $2y = -5x + 3$ and this can be divided by 2 to give $y = -\frac{5}{2}x + \frac{3}{2}$, which is in the form $y = mx + c$.

Gradients of straight-line graphs

Gradients have a sign and a value, for example -0.6 and 4. ← The gradient of a line is the steepness of the line.

The sign of the gradient

The gradient of a straight line can be positive (if y increases as x increases), negative (if y decreases as x increases) or zero (if the value of y stays the same as x increases).

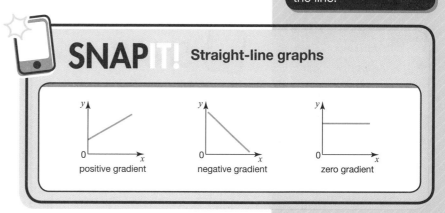

SNAPIT!   **Straight-line graphs**

positive gradient   negative gradient   zero gradient

Finding the gradient

To find the size of the gradient, draw a triangle as shown. Then use this relationship for the gradient:

Gradient, $m = \dfrac{\text{change in } y \text{ values}}{\text{change in } x \text{ values}}$

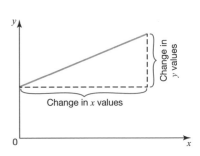

To determine whether the gradient is positive or negative, look at which way the line slopes. Only include the sign if the gradient is negative. If the gradient is positive, say 4, do not write $+4$.

Finding the gradient of a straight line joining two points

SNAP IT!   Gradient of a line

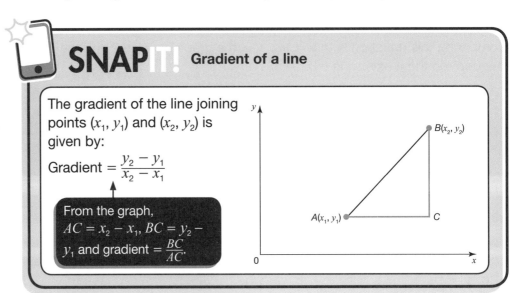

The gradient of the line joining points (x_1, y_1) and (x_2, y_2) is given by:

Gradient $= \dfrac{y_2 - y_1}{x_2 - x_1}$

From the graph, $AC = x_2 - x_1$, $BC = y_2 - y_1$ and gradient $= \dfrac{BC}{AC}$.

WORKIT!

Find the gradient of the straight line joining the points A(−3, 2) and B(1, 6).

$\text{Gradient} = \dfrac{6-2}{1-(-3)} = \dfrac{4}{4} = 1$

Take care with the signs. Make sure that you give the x and y coordinates in the same order.

Finding the midpoint of a straight line

NAIL IT!

You need to remember this formula as it will not be given in the formula booklet.

SNAP IT!

Midpoint of a line

The midpoint of the line joining the points (x_1, y_1) and (x_2, y_2) is given by:

$\left(\dfrac{x_1 + x_2}{2}, \dfrac{y_1 + y_2}{2} \right)$

WORKIT!

Find the midpoint of the line joining the two points with coordinates (2, 6) and (8, 4).

$\left(\dfrac{2+8}{2}, \dfrac{6+4}{2} \right) = (5, 5)$

Finding the equation of a straight line

Given the gradient and y-intercept

Given the gradient of a line, m, and the intercept on the y-axis, c, you can write down the equation of the line.

WORKIT!

Find the equation of the line shown.

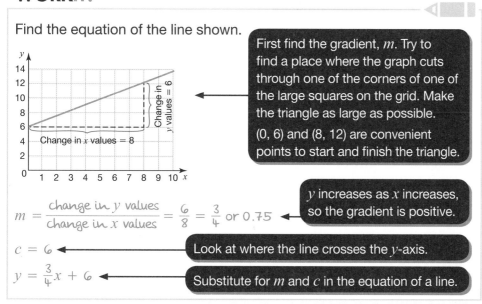

First find the gradient, m. Try to find a place where the graph cuts through one of the corners of one of the large squares on the grid. Make the triangle as large as possible.

(0, 6) and (8, 12) are convenient points to start and finish the triangle.

$m = \dfrac{\text{change in } y \text{ values}}{\text{change in } x \text{ values}} = \dfrac{6}{8} = \dfrac{3}{4}$ or 0.75

y increases as x increases, so the gradient is positive.

$c = 6$ ◄ Look at where the line crosses the y-axis.

$y = \dfrac{3}{4}x + 6$ ◄ Substitute for m and c in the equation of a line.

Given the gradient and the coordinates of a point

SNAPIT!   Equation of a straight line

The equation of a straight line with gradient m and which passes through a point (x_1, y_1) is given by:

$y - y_1 = m(x - x_1)$

NAILIT!

You need to remember the formula for the equation of a straight line as it will not be given in the formula booklet.

WORKIT!

Find the equation of the straight line with a gradient of 2 that passes through the point (2, 5).

$y - 5 = 2(x - 2)$ ◄ Substitute for m, x_1 and y_1.

$y - 5 = 2x - 4$ ◄ Expand the brackets and rearrange into the form $y = mx + c$.

$y = 2x + 1$

This is in the form $y = mx + c$, so you can immediately see that $m = 2$ and c (the intercept on the y-axis) is 1.

Parallel and perpendicular lines

Parallel lines

For two lines to be **parallel** to each other, they must have the same gradient.

For example, the equation of the line that is parallel to the line $y = 3x - 2$ and intersects the y-axis at $y = 4$ is:

$y = 3x + 4$ ← $m = 3$ and $c = 4$

Perpendicular lines

When two lines are **perpendicular** to each other (i.e. they make an angle of 90°), the product (multiple) of their gradients is −1.

If one line has a gradient m_1 and the other a gradient of m_2 then:

$m_1 m_2 = -1$

For example, if a straight line has gradient $-\frac{1}{3}$, then the gradient of the line perpendicular to this is given by:

$\left(-\frac{1}{3}\right) m_2 = -1$ so gradient $m_2 = 3$.

CHECK IT!

1 A straight line has the equation $2y = 4x - 5$.

   a Write down the gradient of the line.

   A line is drawn perpendicular to this line.

   b Write down the gradient of the perpendicular line.

   The perpendicular line crosses the y-axis at (0, 5).

   c Write down the equation of the line.

2 Find the equation of the line that has gradient 3 and passes through the point (2, 3).

3 A line has a gradient of 2 and passes through the point (−1, 0).

   Find the equation of the line in the form $ay + bx + c = 0$.

4 A straight line passes through the points $A(-2, 0)$ and $B(6, 4)$.

   a Find the gradient of the line AB.

   b The midpoint of AB is M. Find the coordinates of M.

   c A straight line is drawn through point M which is perpendicular to the line AB.

      i Write down the gradient of this line.

      ii Find the equation of this line.

Quadratic graphs

Completing the square

Completing the square can be used to:

- solve quadratic equations
- help draw a quadratic graph by finding turning points.

Completing the square: coefficient of $x^2 = 1$

To complete the square for the quadratic $x^2 - 2x + 7$:

> The **coefficient** of x^2 is the number in front of x^2. If the first term is just x^2, the coefficient is 1.

1 Separate the x terms from the constant: $(x^2 - 2x) + 7$.

2 Write the perfect square to give the x^2 and x terms: $(x - 1)^2$.

3 Then subtract a constant to make the identity true: $x^2 - 2x \equiv (x - 1)^2 - 1$.

4 Combine this with the constant in the original quadratic:

$$x^2 - 2x + 7 = (x - 1)^2 - 1 + 7$$
$$= (x - 1)^2 + 6$$

> This example only used integers, but the method also works with fractions.

Completing the square: coefficient of $x^2 = 2$

To complete the square for the quadratic $2x^2 - 6x + 5$:

1 Remove the coefficient of x^2 to outside a bracket: $2\left[x^2 - 3x + \frac{5}{2}\right]$.

2 Use the method above to complete the square for the contents of the square bracket: $2\left[\left(x - \frac{3}{2}\right)^2 - \frac{9}{4} + \frac{5}{2}\right] = 2\left[\left(x - \frac{3}{2}\right)^2 + \frac{1}{4}\right]$.

3 Multiply the expression found by the number outside: $2\left(x - \frac{3}{2}\right)^2 + \frac{1}{2}$

So, $2x^2 - 6x + 5 = 2\left(x - \frac{3}{2}\right)^2 + \frac{1}{2}$.

Solving a quadratic equation by completing the square

Quadratic equations can be solved by completing the square.

WORKIT!

a Show that $x^2 + 6x + 4$ may be expressed in the form $(x + 3)^2 - 5$.

$$x^2 + 6x + 4 = (x + 3)^2 - 9 + 4$$
$$= (x + 3)^2 - 5$$

b Use your answer to part a to solve the quadratic equation $x^2 + 6x + 4 = 0$, giving your answers to 2 decimal places.

$$x^2 + 6x + 4 = 0$$
$$(x + 3)^2 - 5 = 0$$
$$(x + 3)^2 = 5$$
$$x + 3 = \pm\sqrt{5}$$

> You must include both the positive and negative values when you square-root a number, giving two roots (i.e. solutions) rather than just one.

$$x = -3 + \sqrt{5} \text{ or } x = -3 - \sqrt{5}$$
$$x = -0.76 \text{ or } x = -5.24 \text{ (to 2 d.p.)}$$

NAILIT!

You must use completing the square to solve the quadratic equation. You will lose marks if a method is specified in the question and you use a different method.

Sketching a quadratic graph

The shape of the graph of the quadratic equation $y = ax^2 + bx + c$ depends on the sign of a. The curve is ∪-shaped if a is positive and ∩-shaped if a is negative.

Finding the point or points where the curve intersects the x-axis

It is important to note that not all quadratic graphs intersect the x-axis.

'Intersect' means 'to cross'.

To find where the graph intersects the x-axis, solve the equation $ax^2 + bx + c = 0$. These values of x are called the **roots** of the equation. Mark the root(s) of x on the x-axis.

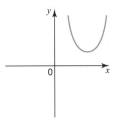

This graph does not intersect the x-axis so there are no roots

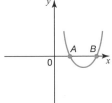

This graph intersects the x-axis in two places so there are two roots.

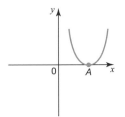

This graph intersects the x-axis in only one place so there is one root.

Finding turning points by completing the square

The **turning points** on a curve are the points where the gradient of the curve is zero. For a quadratic graph this will be a **maximum** point if the curve is ∩-shaped or a **minimum** point if the curve is ∪-shaped.

When the square has been completed, the equation for the curve will look like:

$$y = a(x + p)^2 + q$$

When $x = -p$, the value of the bracket is zero and, since the bracket is squared, this is its minimum value (since it cannot be negative). So the minimum value of y is q.

The turning point will be at $(-p, q)$.

The axis of symmetry will be the line $x = -p$.

For example, for $y = 2(x + 2)^2 - 1$, $a = 2$, $p = 2$ and $q = -1$.

The curve will be ∪-shaped with a minimum point at $(-2, -1)$ and an axis of symmetry of $x = -2$.

The maximum and minimum points of curves are called turning points as the gradient changes sign either side of the point. In this curve the gradient to the left is negative and to the right is positive. The gradient at the turning point itself is zero.

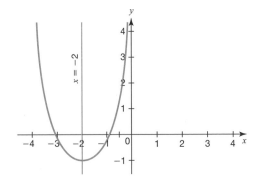

WORKIT!

a Express $4x^2 - 12x + 9$ in the form $a(x + b)^2 + c$.
Give the values of a, b and c.

$$4x^2 - 12x + 9 = 4\left[x^2 - 3x + \frac{9}{4}\right]$$

> Write the coefficient of x^2 as a factor of the expression.

$$= 4\left[\left(x - \frac{3}{2}\right)^2 - \frac{9}{4} + \frac{9}{4}\right]$$

$$= 4\left(x - \frac{3}{2}\right)^2$$

> Compare the expression with $a(x + b)^2 + c$.

$$a = 4, \ b = -\frac{3}{2}, \ c = 0$$

b Hence sketch the graph of $y = 4x^2 - 12x + 9$, including the coordinates of the turning point.

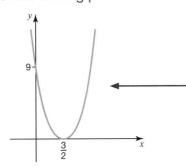

> The coefficient of x^2 is positive, so the graph is ∪-shaped.

> At the minimum point, $x = \frac{3}{2}$. When $x = \frac{3}{2}$, $y = 0$
> The coordinates of the turning point are $\left(\frac{3}{2}, 0\right)$.

WORKIT!

> The graph is ∩-shaped, so the coefficient of x^2 is negative. Use the roots to write the factors of the quadratic.

This is a quadratic graph.

a Write down the roots of the graph.

$x = -2$ and $x = 3$

b Find the equation of the quadratic graph.

$$-(x + 2)(x - 3) = -(x^2 - x - 6)$$

$$= -x^2 + x + 6$$

The equation is $y = -x^2 + x + 6$.

c Use the graph to estimate the solutions to $-x^2 + x + 6 = 2$.

$$x = -1.55 \text{ and } x = 2.55$$

> Estimate the x values when $y = 2$.

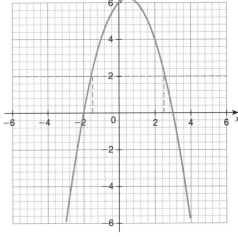

DO IT!

Produce a poster showing the information you need to sketch a quadratic graph.

CHECK**IT!**

1 a Write $2x^2 - 12x + 1$ in the form $a(x + b)^2 + c$, where a, b and c are integers.

b Hence or otherwise, for the graph $y = 2x^2 - 12x + 1$

   i find the coordinates of the turning point

   ii find the roots to 1 decimal place.

c Sketch the graph of $y = 2x^2 - 12x + 1$.

2 Write the equation for each quadratic graph.

a

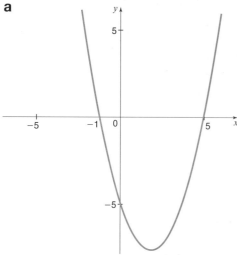

b

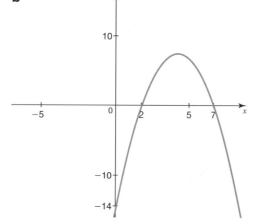

3 a Write $x^2 + 12x - 16$ in the form $(x + a)^2 + b$.

b Write down the coordinates of the turning point of the graph of $y = x^2 + 12x - 16$.

Recognising and sketching graphs of functions

Drawing cubic graphs

Cubic graphs have an x^3 term and have an equation of the form:

$y = ax^3 + bx^2 + cx + d$

where a, b, c and d are constants and where b, c or d could be zero.

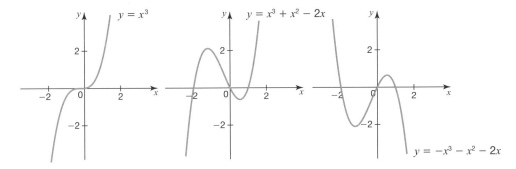

Graph has one root and no turning points.

Graph has two turning points and three roots.

Graph has two turning points and three roots.

WORKIT!

Draw the graph of $y = x^3 + x^2 - 6x$ between $x = -4$ and $x = 3$.

When $x = -4$, $y = (-4)^3 + (-4)^2 - 6(-4) = -24$. ◄————

x	-4	-3	-2	-1	0	1	2	3
$y = x^3 + x^2 - 6x$	-24	0	8	6	0	-4	0	18

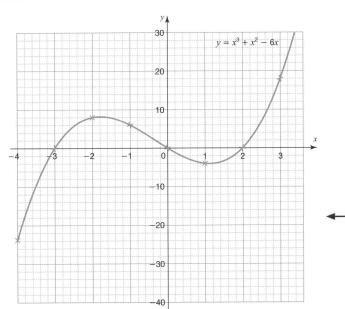

> Draw a table and substitute each value of x into the equation to find the value of y. (The first substitution is shown here.)

> **NAIL**IT!
>
> When substituting into an expression with many terms, it is best to write the calculation down.

> Plot the points in the table and join them with a smooth curve.

Exponential functions are ones in the form $y = a^x$. You will only see exponential functions where a is a positive number.

Recognising sketch graphs

You looked at sketching quadratic graphs in the previous section. You also need to be able to recognise and sketch graphs of the following functions.

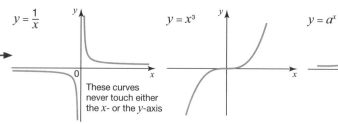

Reciprocal function

$y = \dfrac{1}{x}$

These curves never touch either the x- or the y-axis

Cubic function

$y = x^3$

Exponential function

$y = a^x$

The intercept on the y-axis is 1.

There is more information on the trigonometric ratios on page 156.

The trigonometric functions

You need to know about the three trigonometric functions, **sine**, **cosine** and **tangent** – often abbreviated to sin, cos and tan.

The sine graph, $y = \sin x$

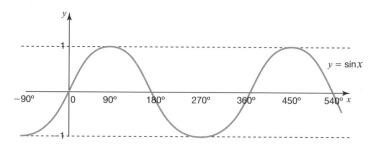

The cosine graph, $y = \cos x$

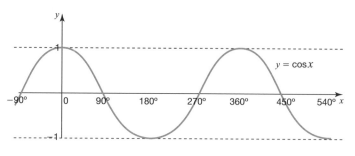

The tangent graph, $y = \tan x$

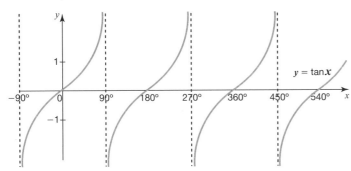

Using the trigonometric graphs to find angles

These graphs can be used to find angles.

WORK**IT**!

1 Find all the values of the angle θ in the range $0° \leq \theta \leq 360°$ satisfying $2\sin\theta = 1$.

$2\sin\theta = 1$

$\sin\theta = \dfrac{1}{2}$

Sin^{-1} means 'angle with a sine of'.

$\theta = \sin^{-1}\left(\dfrac{1}{2}\right)$

$\theta = 30°$

Also, $\theta = 180 - 30 = 150°$

The line $y = 0.5$ (or $y = \dfrac{1}{2}$) lets you read off the values for the angle.

Hence $\theta = 30°,\ 150°$

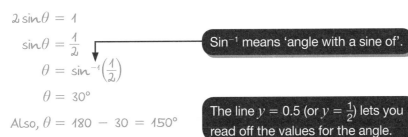

2 The graph shows the curve $y = \sin x$ in the interval $0° \leq x \leq 720°$.

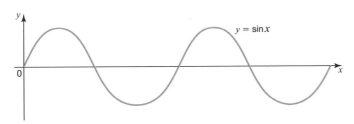

a Write down the coordinates of all the points of intersection with the x-axis.

(0, 0), (180, 0), (360, 0), (540, 0), (720, 0)

b Write down the coordinates of all the turning points for this graph.

(90, 1), (270, −1), (450, 1), (630, −1)

DO**IT**!

Type an equation into an internet search engine and find out what the graph looks like.

Make a small change to the equation and search again to see how this changes the graph.

To produce the graph $y = x^3$, you need to enter the equation as $y = x\ ^\wedge 3$ as the $^\wedge$ symbol is used to indicate a power or **exponential**.

CHECK**IT!**

1 Here are the equations of six graphs.

a $y = 5x^3$

b $y = \cos x$

c $y = 3^x$

d $y = x^2 - 4x + 4$

e $y = \sin x$

f $y = \frac{2}{x}$

Match each equation to one of the graphs A to F.

A
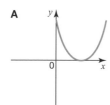

B
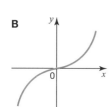

C
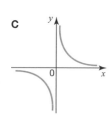

D
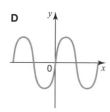

E
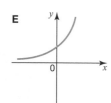

F
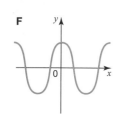

2 a Sketch the graph of $y = \tan x$ for $0° \leq x \leq 360°$.

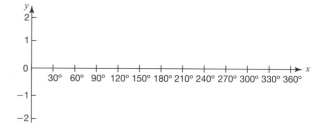

b One solution to $\tan x = \sqrt{3}$ is $x = 60°$.

Find another solution for x in the range $0° \leq x \leq 360°$.

3 Here are the equations of four graphs. Match each equation to one of the graphs A to G.

a $y = 1 - x$

b $y = 2x - 1$

c $y = x^2 - 3$

d $y = 3 - x^2$

A
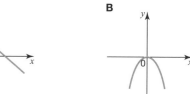

B

C
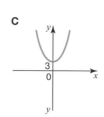

D
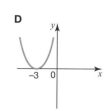

E
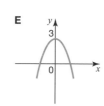

F
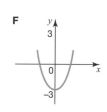

G
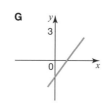

Translations and reflections of functions

Translating the graph of $y = f(x)$

A **translation** is a vertical movement, a horizontal movement or both. If you are given a graph of a function in the form $y = f(x)$, then the graph of a new function may be obtained from the original graph by applying a translation.

$y = f(x)$ translated to $y = f(x + a)$ is a **translation** of $-a$ units **parallel to the x-axis**. The whole curve moves a units to the left. Similarly, $y = f(x - b)$ is a translation of b units to the right.

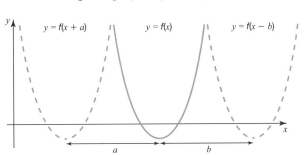

$y = f(x)$ translated to $y = f(x) + a$ is a **translation** of a units **parallel to the y-axis**. If a is positive, the whole graph moves up a units and if a is negative it moves down by a units.

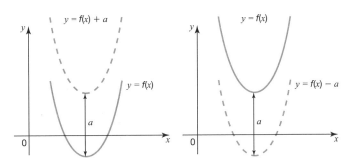

Reflecting the graph of $y = f(x)$

$y = -f(x)$ is a **reflection** of $y = f(x)$ **in the x-axis**.

$y = f(-x)$ is a **reflection** of $y = f(x)$ **in the y-axis**.

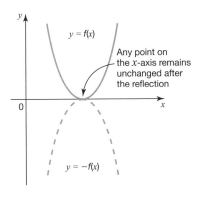

Any point on the x-axis remains unchanged after the reflection

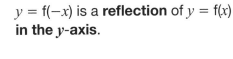
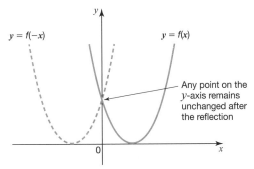

Any point on the y-axis remains unchanged after the reflection

WORKIT!

1   The diagram shows a sketch of the graph $y = f(x)$. The graph passes through the points (1, 0) and (5, 0) and has a turning point at (3, −4).

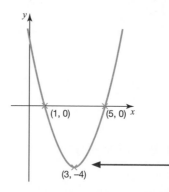

A turning point is a point on the curve where the gradient is zero. This will occur at any maximum or minimum points.

Sketch the following graphs, using a separate set of axes for each graph. In each case, indicate the coordinates of the turning point and the coordinates of the points of intersection of the graph with the x-axis.

a $y = f(x + 1)$

b $y = -f(x)$

$y = f(x + 1)$ is a translation of $y = f(x)$ by one unit to the left parallel to the x-axis).

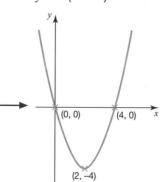

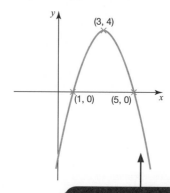

The negative sign is before $f(x)$, so this is a reflection in the x-axis.

Combinations of reflections and translations

Translations and reflections can be combined. The order in which they are carried out does not matter.

Here are some combinations of reflections and translations for this graph of $y = f(x)$.

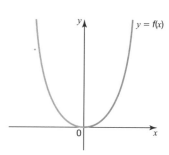

NAILIT!

Remember to add the coordinates of the points of intersection with the x-axis and of the turning point. You would lose marks if you only put the x coordinates of these points on the x-axis.

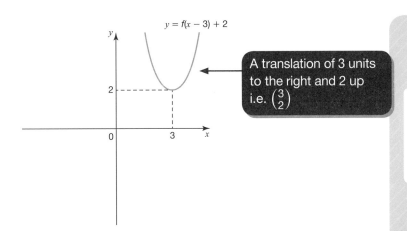

A translation of 3 units to the right and 2 up i.e. $\begin{pmatrix} 3 \\ 2 \end{pmatrix}$

DO IT!

Draw a poster summarising translations and transformations of functions.

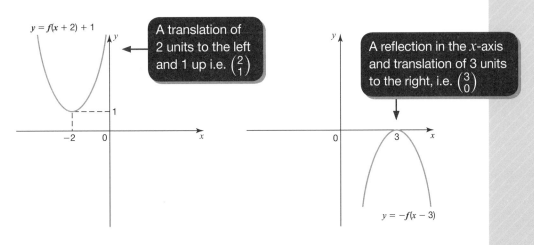

A translation of 2 units to the left and 1 up i.e. $\begin{pmatrix} 2 \\ 1 \end{pmatrix}$

A reflection in the x-axis and translation of 3 units to the right, i.e. $\begin{pmatrix} 3 \\ 0 \end{pmatrix}$

CHECK IT!

1 The diagram shows the graph of $y = f(x)$. The graph has a turning point at $(2, 5)$ and intersects the x-axis at the points $(-2, 0)$ and $(6, 0)$.

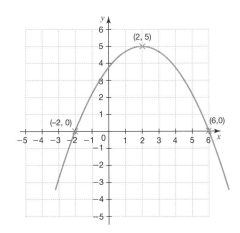

Write down the coordinates of the turning point of the curve with equation

a $y = f(x - 1)$ **c** $y = -f(x)$

b $y = f(x + 3)$ **d** $y = f(-x)$

2 The diagram shows the graph of $y = f(x)$. The graph has a maximum point at $(1, 2)$.

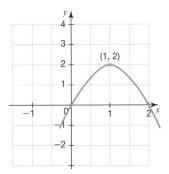

Sketch the following graphs, using a separate set of axes for each graph. Mark the coordinates of the turning point on each graph.

a $y = f(x - 1)$ **c** $y = -f(x - 1)$

b $y = f(x) + 2$ **d** $y = f(x - 1) + 2$

Equation of a circle and tangent to a circle

Equation of a circle

SNAP IT!   Equation of a circle

The equation of a circle with centre the origin (0, 0) and radius r is:

$$x^2 + y^2 = r^2$$

So the equation $x^2 + y^2 = 25$ represents the equation of circle with centre at the origin and with a radius of 5.

WORKIT!

A circle has the equation $x^2 + y^2 = 9$.

a Write down the coordinates of the centre of the circle and give the radius of the circle.

Centre is at (0, 0).

Radius = 3

> Comparing with $x^2 + y^2 = r^2$, $r^2 = 9$.

b Prove that the point P $(\sqrt{3}, \sqrt{6})$ lies on the circle.

If these points lie on the circle, then:

$$x^2 + y^2 = (\sqrt{3})^2 + (\sqrt{6})^2$$

> If the point lies on the circle its coordinates will satisfy the equation of the circle.

$$= 9$$

This is the same as the right-hand side of the equation of the circle. The point therefore satisfies the equation and so lies on the circle.

Equation of a tangent to a circle

The tangent to a circle is perpendicular to the radius. Using the coordinates of where the tangent touches the circle and the coordinates of the centre of the circle, find the gradient of the radius joining these two points using the formula:

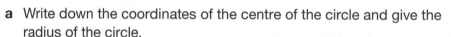

$$\text{Gradient} = \frac{y_2 - y_1}{x_2 - x_1}$$

Use this gradient to work out the gradient of the tangent, as these two lines are perpendicular to each other. If one line has a gradient m_1 and the other a gradient of m_2 then $m_1 m_2 = -1$.

Use the coordinates of the point where the tangent touches the circle and the gradient of the tangent and substitute them into the following formula to give the equation of the tangent:

$$y - y_1 = m(x - x_1)$$

> You learned how to find the equation of a line from the gradient and a point on the line on page 77.

WORKIT!

A circle with equation $x^2 + y^2 = 25$ has centre O and a tangent to the circle at the point $P(3, 4)$.

a Find the gradient of the line OP.

Gradient of $OP = \dfrac{y_2 - y_1}{x_2 - x_1} = \dfrac{4 - 0}{3 - 0} = \dfrac{4}{3}$

b Using your answer to part a, write down the gradient of the tangent to the circle at point P.

As radius OP and tangent are perpendicular the product of their gradients is -1.

Gradient of tangent $= -\dfrac{3}{4}$

c Find the equation of the tangent passing through point P.

$y - y_1 = m(x - x_1)$

$y - 4 = -\dfrac{3}{4}(x - 3)$

> Use this with $m = -\dfrac{3}{4}$ and $(x_1, y_1) = (3, 4)$.

$4y - 16 = -3x + 9$

$4y = -3x + 25$

DO IT!

Draw a flowchart for finding the equation of the tangent to a circle.

> The easiest way to find the gradient of the tangent is to invert the gradient of OP (put the fraction upside down) and change the sign.

✓ CHECKIT!

1 A circle has the equation $x^2 + y^2 = 49$.

  a Write down the coordinates of the centre of the circle.

  b Write down the radius of the circle.

2 The diagram shows a circle with centre at the origin O.
  a Find the equation of the circle.

  A tangent to the circle is drawn at $(8, 6)$.

  b Write down the gradient of the tangent at this point.

  c Work out the equation of the tangent at $(8, 6)$.

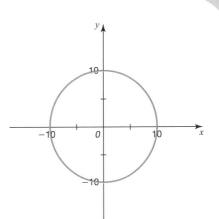

Real-life graphs

Distance–time graphs

These are the main features of any **distance–time graph**.

- Gradient of a line is the average speed. A steeper line means a higher speed.
- Horizontal sections mean the motion has stopped.
- A negative gradient shows the speed of the return journey.

Below is a distance–time graph for someone's journey to and from a meeting.

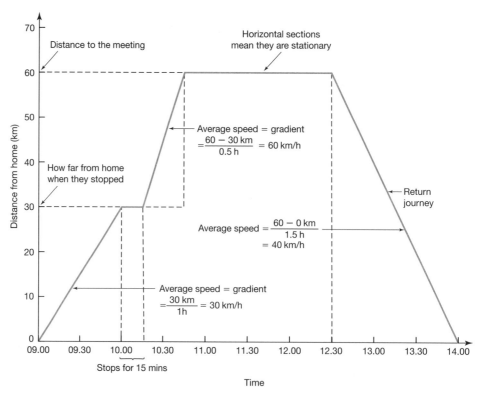

There are lots of things you can tell from this graph and some of them are marked on it. Other things you can tell include:

- The average speed in the first hour is lower than the average speed when they continued after stopping for 15 minutes.
- They stopped at 10 am and waited for 15 minutes.
- The meeting started at 10.45 am and finished at 12.30 pm so it took $1\frac{3}{4}$ hours.

Velocity–time graphs

These are the main features of a **velocity–time graph**.

- The gradient represents the acceleration. A positive gradient represents a positive acceleration and a negative gradient represents a negative acceleration (or deceleration). The units for acceleration are m/s^2.
- Horizontal sections represent constant velocity.
- The area under the graph represents the distance travelled.

DO IT!

Draw and annotate a distance–time graph of a journey to and from a friend's house.

STRETCH**IT!**

If the velocity–time graph is a curve, work out the distance travelled by dividing the area into strips. Find the area of each strip and then add them up. Alternatively, you can count the squares and multiply this number by the distance that is equivalent to one square.

If the graph is a curve, you can still find the acceleration. Finding the gradient of a curve is covered on page 121.

WORK**IT!**

This velocity–time graph is for a car moving along a straight horizontal road. Its initial velocity is 20 m/s.

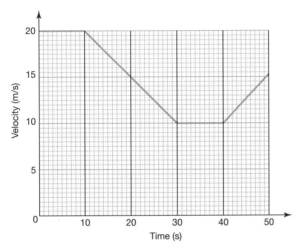

> Area of trapezium $= \frac{1}{2}$ (sum of the two parallel sides) × distance between them. Alternatively, divide the shape into a triangle and a rectangle and add the two areas together. Note you will not be given the formula for the area of a trapezium.

a Find the distance travelled while the car is decelerating.

Distance = area of trapezium ←

$= \frac{1}{2}(20 + 10) \times 20$ ←

> Base of trapezium is when the car is decelerating, between 10 and 30 seconds.

$= 300\,m$

> Sides of trapezium are the velocity before decelerating and the velocity after.

b Find the total distance travelled while travelling at constant speed.

Total distance while travelling at constant speed $= 20 \times 10 + 10 \times 10 = 300\,m$ ←

> Find areas of the rectangles where the car is travelling at constant speed.

c During the last stage of the motion the car accelerates. Calculate the acceleration.

Acceleration = gradient $= \dfrac{15 - 10}{10} = 0.5\,m/s^2$

d Calculate the total distance travelled during the motion described by the graph.

Distance travelled while accelerating $= \frac{1}{2}(10 + 15) \times 10 = 125\,m$

From a, b and d, total distance travelled $= 300 + 300 + 125$

$= 725\,m$

> The total distance is the total area under the velocity–time graph.

Graphs in finance

Graphs are used in finance to show costs.

This graph shows how the cost of water varies with the volume used.

These are the key features of the graph:

- The intercept on the y-axis shows that there is a fixed cost when no water is used.

- Reading off the graph, the fixed charge is £34.

- The gradient of the graph is the cost per m³ of the water. The line goes through (50, 160) and (0, 34) so these points can be used to work out the cost per m³.

  Cost per m³ $= \dfrac{160 - 34}{50 - 0} = \dfrac{126}{50} =$ £2.52

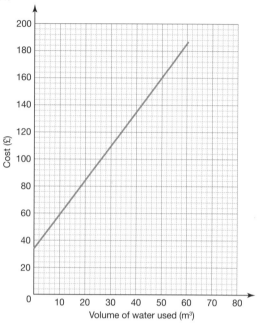

✓ CHECKIT!

1 The distance–time graph represents a cycle ride.

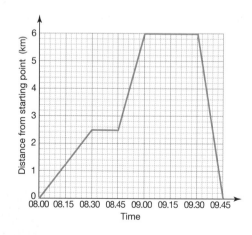

a Find the average speed for the first half-hour.

b At 08.30, the cyclists stopped for a break. How long was this break in hours?

c Find the average speed for the return journey.

2 A particle starts from rest and travels in a straight line. It accelerates uniformly for 2 s, then travels at a constant velocity of u m/s for 10 s, before decelerating uniformly to rest in 3 s. The total distance travelled by the particle is 50 m.

a Draw a velocity–time graph to show the motion of the particle.

b Find the value of u.

c Find the magnitude of the deceleration.

Generating sequences

Arithmetic and geometric sequences

An **arithmetic sequence** (sometimes called a linear sequence or an arithmetic progression) is a sequence of numbers with a constant difference (positive or negative) between each number and the one before. Add that same amount to give the next number in the sequence.

1, 3, 5, 7,…	Add 2 to get the next term.
13, 18, 23, 28,…	Add 5 to get the next term.
10, 6, 2, −2, −6,…	Add −4 (i.e. subtract 4) to get the next term.

WORKIT!

The first four terms of a sequence are 33, 30, 27, 24,…

a Write down the next two terms.

  21, 18

b What is the rule for continuing this sequence?

  Subtract 3

  The sequence goes down by 3 each time.

c What is the first negative number in this sequence?

  18, 15, 12, 9, 6, 3, 0, −3

  The first negative term is −3.

A **geometric sequence** (sometimes called a geometric progression) is a sequence of numbers with a constant ratio between each number and the one before. Multiply each number in the sequence by the same amount to give the next number in the sequence.

1, 3, 9, 27, 81,…	Multiply by 3 to get the next term.
$\frac{1}{2}, \frac{1}{4}, \frac{1}{8}, \frac{1}{16},\ldots$	Multiply by $\frac{1}{2}$ to get the next term.
1, $\sqrt{3}$, 3, $3\sqrt{3}$,…	Multiply by $\sqrt{3}$ to get the next term.

DOIT!

Summarise the difference between an arithmetic and a geometric sequence in 21 words.

Other common sequences

A Fibonacci sequence usually starts 1, 1, but can start with any two numbers.

SNAPIT!   Common sequences

Sequence	Name	Rule
1, 4, 9, 16, 25, 36,…	Square numbers	Square the number of each term
1, 8, 27, 64, 125, 216,…	Cube numbers	Cube the number of each term
1, 3, 6, 10, 15, 21,…	Triangular numbers	As shown in the diagram on page 96
1, 1, 2, 3, 5, 8,…	Fibonacci numbers	Add together the previous two terms

Triangular numbers

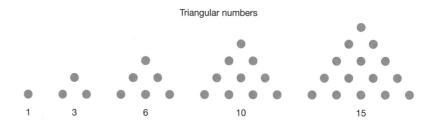

1   3   6   10   15

NAILIT!

Remember these common sequences, as they often come up in exam questions.

WORKIT!

Write down the next two terms in these sequences.

A geometric sequence with a ratio of 3

a $\frac{1}{9}, \frac{1}{3}, 1, 3,...$

    9, 27

b 7, 5, 3, 1,...

An arithmetic sequence with a common difference of −2

    −1, −3

c 1, 8, 27,...

    64, 125   ◄   The cube numbers

d 1, 1, 2, 3, 5,...

    8, 13   ◄   Fibonacci sequence

NAILIT!

Sometimes you will be given the rule and a term, and be asked to find the previous term. You need to use the inverse operations from the rule: for example, 'multiply by 2 and subtract 3' would become 'add 3 and divide by 2'. Note that you need to do the opposite operations and in the **opposite order**.

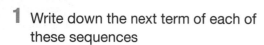

CHECKIT!

1 Write down the next term of each of these sequences

   a 5, 8, 11, 14,...

   b 2.2, 2.4, 2.6, 2.8,...

   c 0, −3, −6, −9,...

   d 2, 12, 72,...

   e $\frac{1}{6}, \frac{1}{12}, \frac{1}{24},...$

   f $\frac{1}{2}, -\frac{1}{4}, \frac{1}{8},...$

2 The term-to-term rule for a sequence is square and add one. The first term of the sequence is −4. Find the 2nd and 3rd terms.

3 The term-to-term rule for a sequence is double and add one. The third term of the sequence is 12. Find the 1st and 2nd terms.

The *n*th term

The *n*th term of a sequence is a formula that gives the value of any term depending on its position in the sequence. For example, to find the 3rd term, put 3 into the formula.

The *n*th term of a linear sequence

To find the *n*th term of the linear sequence: 8, 13, 18, 23,…

n	1	2	3	4
Term	8	13	18	23
Difference		5	5	5

> Write the position of the term above the term.

> Find the difference between terms.

As there is a constant difference of 5 the first part of the formula will be 5*n*.

5*n*	5	10	15	20
Term − 5*n*	3	3	3	3

> Work out the difference between each term and 5*n*.

The *n*th term is 5*n* + 3.

> Sometimes you may have to subtract a value instead of add.

WORKIT!

Here are the first four terms of a sequence 2, 5, 8, 11,…

a Write an expression, in terms of *n*, for the *n*th term of the sequence.

n	1	2	3	4
Term	2	5	8	11
Difference		3	3	3
3*n*	3	6	9	12
Term − 3*n*	−1	−1	−1	−1

*n*th term = 3*n* − 1

b Find the 10th term in the sequence.

10th term = 3 × 10 − 1 = 29

> Substitute *n* = 10 into the formula for the 10th term.

c Is 200 a term of this sequence? Explain your answer.

3*n* − 1 = 200

3*n* = 201

n = 67

> Put the *n*th term formula equal to the term. If *n* works out as an integer, it is a number in the sequence.

As *n* is a whole number, 200 is a term in the sequence.

The *n*th term of a quadratic sequence

In a quadratic sequence the *n*th term includes a term in n^2.

To find the *n*th term for the sequence 4, 9, 16, 25,… first find the first and second differences between the terms:

Term		4		9		16		25
First difference			5		7		9	
Second difference				2		2		

> The second difference is the difference between successive first differences.

A constant second difference means that there must be an n^2 term. To find the number in front of n^2, divide the second difference by 2 (in this case, $2 \div 2 = 1$).

So the formula starts n^2. So we can summarise what we know:

n	1	2	3	4
Term	4	9	16	25
n^2	1	4	9	16
Term $- n^2$	3	5	7	9

> We know each term contains n^2. So work out the difference between each term and n^2.

Use the differences between this set of terms to work out the linear part of the sequence (the term in *n*).

Difference		2		2		2	
$2n$	2		4		6		8
Term $- n^2 - 2n$	1		1		1		1

> A difference of 2 means the linear part will start with $2n$.

Combining the terms gives *n*th term $= n^2 + 2n + 1$

DO IT!

Draw a flowchart for finding the *n*th term of a sequence. The chart needs to work with both arithmetic and quadratic sequences.

✓ CHECKIT!

1 The *n*th term of a sequence is $50 - 3n$.

  a Write down the first three terms of this sequence.

  b Does the sequence contain the number 34?

  c Work out the value of the first term of this sequence that is negative.

2 The *n*th term of a sequence is given by 2×3^n.

  a Write down the first four terms of this sequence.

  b All the terms in this sequence are multiples of 6. Explain why.

3 The first four terms of an arithmetic sequence are -1, 1, 3, 5.

  a Work out an expression for the *n*th term of this sequence.

  b The *x*th term of this sequence has a value of 59. Find the value of *x*.

4 A sequence has the following first four terms:

4, 17, 38, 67

Find a formula for the *n*th term of this sequence.

Arguments and proofs

Arguments and proofs can crop up in almost any topic. As a result of this there is no set way of tackling them. The main thing to remember is that you have to sometimes put letters for values, angles, sides, and so on. Sometimes there are several ways to solve the problem.

WORKIT!

Here are four expressions

$x - 1$ $3x - 3$ $x^2 - 4$ $2(x + 1)$

x is a positive integer. Explain which expression has a value which is

a always even

    2 is a factor of $2(x + 1)$.

    So $2(x + 1)$ is always even.

b always a multiple of 3

    $3x - 3 = 3(x - 1)$, so 3 is a factor.

    So $3(x - 1)$ is always a multiple of 3.

c never zero.

    When $x = 1$, $x - 1 = 0$

    When $x = 1$, $3x - 3 = 0$

    When $x = 2$, $x^2 - 4 = 0$ ◄—— Sometimes you can show something is **not** true by giving one **counter-example.**

    $2(x + 1)$ is always greater than zero, because x is always greater than 0.

    So $2(x + 1)$ is never zero.

WORKIT!

Prove that $(2x - 1)^2 - (x - 2)^2$ is a multiple of 3 for all integer ◄—— Expand the brackets.
values of x.

$$(2x - 1)^2 - (x - 2)^2 = 4x^2 - 4x + 1 - (x^2 - 4x + 4)$$
$$= 4x^2 - 4x + 1 - x^2 + 4x - 4$$
$$= 3x^2 - 3$$
$$= 3(x^2 - 1) \quad ◄\text{—— Factorise the expression.}$$

The 3 outside the brackets shows that the result is a multiple of 3 for all integer values of x.

NAIL IT!

If you are asked to prove something for even and/or odd numbers, use an expression of the form $2n$ for even numbers and $2n + 1$ for odd numbers.

WORKIT!

Prove that the product of two odd numbers always gives an answer that is odd.

Let two odd numbers be $2m + 1$ and $2n + 1$. ◄─── $2n$ must be even, so $2n + 1$ must be odd.

$(2m + 1)(2n + 1) = 4mn + 2m + 2n + 1$

$\qquad\qquad\qquad = 2(2mn + m + n) + 1$

$2(2mn + m + n)$ must be even, so $2(2mn + m + n) + 1$ must be odd.

Therefore the product of two odd numbers is always an odd number.

Multiply the two expressions to find the product.

Always state what you have been asked to prove at the end of your proof.

✓ CHECKIT!

1 Prove whether each of these statements is true or false.

    a If n is a positive integer, $2n + 1$ is always an odd number.

    b If $x^2 - 9 = 0$, the only value x can be is 3.

    c If $n > 0$, n^2 is always an integer.

    d If $n > 0$, n^2 is always greater than n.

2 Prove that the sum of four consecutive numbers is always even.

3 Prove that the sum of three consecutive integers is always a multiple of 3.

4 a and b are both integers and $a > b$. Prove whether each of these statements is true or false.

    a $\dfrac{a}{b} < 1$

    b $b^2 > a^2$

    c $\sqrt{a} > 1$

1 Multiply out the brackets and simplify where possible

   a $-3(3x - 4)$

   b $4x + 3(x + 2) - (x + 2)$

   c $(x + 3)(2x - 1)(3x + 5)$

2 **a** Factorise the expression $2x^2 + 7x - 4$.

   b Solve the equation $2x^2 + 7x - 4 = 0$.

3 Simplify

   a $(2x^2y)^3$ **c** $\frac{15a^3b}{3a^3b^2}$

   b $2x^{-3} \times 3x^4$

4 Solve the simultaneous equations

   $3x + 2y = 8$

   $5x + y = 11$

5 **a** Show that $\frac{3}{x + 7} = \frac{2 - x}{x + 1}$ can be written

      as $x^2 + 8x - 11 = 0$.

   b Hence solve the equation $\frac{3}{x + 7} = \frac{2 - x}{x + 1}$,

      giving your answers to 2 decimal places.

6 Make x the subject of $\frac{3y - x}{z} = ax + 2$.

7 The function f is such that $f(x) = \frac{x}{3} + 5$.

   a Find $f^{-1}(x)$.

   b The function g is such that $g(x) = 2x^2 + k$, where k is a constant.

      Given that $fg(2) = 10$, work out the value of k.

8 The nth term of a sequence is $30 - 4n$.

   a Write down the first three terms of this sequence.

   b Work out the value of the first term of this sequence that is negative.

9 A circle has the equation $x^2 + y^2 = 21$. Determine whether the point (4, 3) lies inside or outside this circle.

10 Simplify fully $(\sqrt{x} + \sqrt{9y})(\sqrt{x} - 3\sqrt{y})$.

11 **a** Write $2x^2 + 8x + 1$ in the form $a(x + b)^2 + c$, where a, b and c are integers.

   b Hence or otherwise, for the graph $y = 2x^2 + 8x + 1$:

      i find the coordinates of the turning point

      ii find the roots to 1 decimal place.

   c Sketch the graph of $y = 2x^2 + 8x + 1$.

12 The perimeters of rectangle $ABCD$ and triangle EFG are the same.

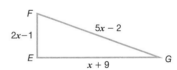

All measurements are in centimetres.

Work out the area of the triangle.

13 $A(-5, 2)$, $B(-2, -2)$, $C(2, 1)$ and $D(-1, k)$ are the vertices of a square.

Find the equation of the diagonal BD.

14 Solve algebraically the simultaneous equations

   $x^2 + y^2 = 4$

   $2y - x = 2$

Introduction to ratios

Writing ratios

A **ratio** is a comparison between quantities. For example, the ratio of two quantities a and b can be written as $a:b$ or as the fraction $\frac{a}{b}$.

> Ratios are like fractions – cancel them if possible.

Cancel ratios such as $32:48$ by dividing both numbers by the highest number that divides exactly into both. Dividing by 16 simplifies the ratio to $2:3$.

> This is the highest common factor.

The quantities must be in the same units before the ratio is cancelled. If the ratios are in different units (e.g. $1\,\text{kg}:200\,\text{g}$):

1 Convert both to the smaller unit: $1000\,\text{g}:200\,\text{g}$.

2 Remove the units: $1000:200$.

3 Cancel the resulting ratio if possible: $5:1$.

WORKIT!

1 On a map a distance of 2 cm represents 5 km. Write the ratio $2\,\text{cm}:5\,\text{km}$ in the form $1:n$.

> First make the units the same.

$$2\,\text{cm}:5\,\text{km} = 2:500\,000 = 1:250\,000$$

2 Reduce the ratio $34:6$ to the form $1:n$ where n is a number correct to 2 significant figures.

> To change the 34 into 1, divide it by 34.
> To keep the ratio the same, divide the other side by 34 too.
> Round the value of n to to 2 s.f.

$$34:6 = \frac{34}{34}:\frac{6}{34} = 1:0.1764\ldots = 1:0.18$$

Dividing a quantity into two parts in a certain ratio

SNAPIT! Dividing a quantity in a certain ratio

To divide a quantity in a certain ratio:

1 Add the numbers in the ratio together.

2 Divide this number into the quantity you are dividing up. This gives you what one part represents.

3 Multiply the one part by each number in the ratio in turn.

4 Check that the numbers you get add up to give the original quantity.

NAILIT!

Always check to see if you have to give the value of all the parts or just one of the parts. Always give the answer in the form requested by the question.

WORKIT!

1 £180 is divided in the ratio of 4 : 5. Find the value of each portion.

Total number of parts = 4 + 5 = 9

1 part = $\frac{£180}{9}$ = £20

So the portions are 4 × 20 = £80 and 5 × 20 = £100 ◄

> Check your answer by adding together the quantities: 80 + 100 = 180.

2 Divide 250 kg in the ratio 0.4 : 2.8.

0.4 : 2.8 = 4 : 28 = 1 : 7 ◄

> Convert the ratio to whole numbers by multiplying by 10 and then cancelling.

Total number of parts = 1 + 7 = 8

1 part = $\frac{250}{8}$ = 31.25 kg

So the portions are 1 × 31.25 = 31.25 kg and 7 × 31.25 = 218.75 kg

Finding the total being divided in a certain ratio

Sometimes you are given one of the values in the ratio and asked to find the total.

NAILIT!

> The method is the same for a three-part ratio. Dividing a quantity in the ratio 2 : 5 : 7, the total number of parts is 2 + 5 + 7 = 14.

WORKIT!

1 The ratio of the number of boys to girls in a choir is 4 : 5. There are 16 boys in the choir. How many children are in the choir?

4 parts = 16 children so 1 part = 4 children

Total number of parts 4 + 5 = 9 parts

Total number of children = 9 × 4 = 36

2 A bag contains red and blue balls. The ratio of red balls to blue balls is 2 : 3. There are ten more blue balls than red balls. Calculate the total number of balls in the bag. ◄

> You need to use algebra to solve this question, as you do not know how many red or blue balls there are.

Let number of red balls = x, so the number of blue balls = $x + 10$

$\frac{\text{number of red balls}}{\text{number of blue balls}} = \frac{2}{3}$ ◄

> Remember that ratios can also be written as fractions.

$\frac{x}{x + 10} = \frac{2}{3}$

$3x = 2(x + 10)$

$3x = 2x + 20$

$x = 20$

Total number of balls = $x + x + 10 = 50$

3 In a vet's practice, the ratio of the number of dogs seen to the number of cats seen in a week is 5 : 3.

50% of the dogs are over the age of 5.

60% of the cats are over the age of 5. ◄

> This is a more challenging question. With some questions you have to put in your own numbers.

What percentages of all the dogs and all the cats seen are over the age of 5?

Total number of parts = 5 + 3 = 8

Assume there are 80 dogs and cats in total. ← Pick a number that is divisible by 8, e.g. 80.

Dividing this in the ratio 5:3 will give 50 dogs and 30 cats.

Number of dogs over 5 years old = 50% of 50 = 25

Number of cats over 5 years old = 60% of 30 = 18

Number of dogs and cats over 5 years old = 25 + 18 = 43

Percentage of dogs and cats over 5 years old = $\frac{43}{80} \times 100 = 53.75\%$

DOIT!

Produce a revision card for each type of ratio problem described here.

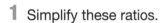

CHECKIT!

1 Simplify these ratios.

  a 2:6

  b 25:60

  c 1.6:3.6

2 Express these ratios as simply as possible, without units.

  a 250 g : 2 kg

  b 25 m : 250 mm

  c 2 cl : 1 l

3 Share £400 in the ratio 3.5 : 2.1.

4 The ratio of the number of girls to the number of boys in a gym is 4 : 3. There are 180 girls in the gym. Calculate how many children belong to the gym.

5 Three daughters are aged 21, 25 and 29 years. They are left £150 000 in a will to be divided between them in the ratio of their ages.

Work out how much money the youngest daughter will receive.

6 The number of male and female guests staying at a hotel is in the ratio 5 : 2.

40% of the male guests are over the age of 40.

30% of the female guests are over the age of 40.

Work out the percentage of all guests under the age of 40.

7 A money box contains 10 p and 20 p coins. The ratio of 10 p coins to 20 p coins is 5 : 7. There are six more 20 p coins than 10 p coins.

Calculate the total amount in the money box.

8 In a box of marbles, there are:

• two times as many blue marbles as red marbles

• five times as many red marbles as yellow marbles.

Work out the ratio of blue marbles to red marbles to yellow marbles.

Scale diagrams and maps

On a scale diagram or a map, all the dimensions have been reduced by the same proportion.

For example, a scale drawing of a garden design with a scale of 1:50 means that a distance of 1 in any unit on the drawing represents an actual distance 50 times larger. So a pond with a diameter of 5 cm on the drawing would have an actual diameter of 50 × 5 cm = 250 cm (or 2.5 m).

NAILIT!

Always change any measurements into the same units.

WORKIT!

A map has a scale of 1:50 000. A distance on the map between two villages is 6 cm. What is the actual distance? Give your answer in km.

1 cm on the map is 50 000 cm actual distance.

6 cm on the map is 6 × 50 000 = 300 000 cm actual distance.

$300\,000\,cm = \frac{300\,000}{100}\,m = 3000\,m$ ← Convert 300 000 cm to km.

3000 m = 3 km

Actual distance = 3 km

NAILIT!

Make sure your answer is clearly stated at the end. Sometimes it is best to write a sentence.

WORKIT!

An artist is producing a scale drawing of a horse that is 1.75 m tall. The artist has decided to use a scale of 1:10. How tall will the horse be on the drawing? Give your answer in centimetres.

1.75 m = 175 cm

Height of horse on drawing = $\frac{1}{10}$ × 175 = 17.5 cm ← The scale of 1:10 means that the height of the horse needs to be $\frac{1}{10}$ of the horse's actual height.

DOIT!

Work out the actual distance represented by 4 cm on maps at scales 1:2500, 1:25 000 and 1:500 000.

✓ CHECKIT!

1 A map has a scale of 1:500 000. The actual distance between two cities is 150 km.

How far apart are the cities on the map? Give your answer in cm.

2 The scale diagram shows an island with a port and two ships A and B out at sea.

a The actual distance between ship A and the port is 10 km. Calculate the scale of the map.

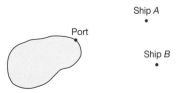

b Work out the actual distance between ships A and B.

Percentage problems

One quantity as a percentage of another

First express the quantities as a ratio in fraction form and then convert this to a percentage by multiplying by 100.

SNAPIT!

A as a percentage of B

Quantity A expressed as a percentage of a quantity B is:

$\frac{A}{B} \times 100\%$

WORKIT!

The theoretical yield of a certain product in a reaction is 0.458 g. The actual yield in an experiment was 0.412 g. Work out the actual yield as a percentage of the theoretical yield. Give your answer to 3 significant figures.

$\frac{actual\ yield}{theoretical\ yield} \times 100 = \frac{0.412}{0.458} \times 100$

$= 89.9563\%$

$= 90.0\%\ (to\ 3\ s.f.)$

> The actual yield is the numerator and the theoretical yield the denominator. Multiply the fraction by 100 to convert it to a percentage.

Percentage change

Percentage change is worked out in a similar way:

Percentage change $= \frac{change}{original\ value} \times 100$

SNAPIT! **Percentage increase/decrease**

The change can be an increase or a decrease, so:

Percentage increase $= \frac{increase}{original\ value} \times 100$

Percentage decrease $= \frac{decrease}{original\ value} \times 100$

> Increase = final value − initial value
> Decrease = initial value − final value

WORKIT!

After 1 year a new car originally costing £15 000 is worth £12 000. What is the percentage decrease?

Decrease = initial value − final value = 15 000 − 12 000 = £3000

Percentage decrease $= \frac{decrease}{original\ value} \times 100 = \frac{3000}{15000} \times 100 = 20\%$

Adding a percentage onto a quantity

In many situations a certain percentage is added onto a quantity. For example, the price of an article could increase by a certain percentage or you may need to find the price of an article after the VAT has been added on.

> You could work out 20% of 13500 and then add this on, but there is a quicker way if you can use a calculator.

WORKIT!

1 A car costs £13500 before VAT. Find the cost of the car after VAT of 20% has been added.

> You are adding 20% to the original amount (i.e. 100%), so you need to find 120% of the original amount, which is the same as multiplying by 1.2.

> Work out the multiplier.

$$120\% = \frac{120}{100} = 1.2$$

> Multiply the original cost by the multiplier.

Price of car including VAT at 20% = 13500 × 1.2 = £16200

2 A company's sales in one year were £21200. The sales the next year increased by 140%. Work out the sales in the second year.

> 100% + 140% = 240%.

Multiplier = 240% = $\frac{240}{100}$ = 2.4

Second year's sales = 21200 × 2.4 = £50880

Subtracting a percentage from a quantity

Some things go down in price. For example, the value of most cars goes down with age (called depreciation); goods in a sale are often reduced by a certain percentage.

WORKIT!

In a sale, all goods are reduced by 35%. The original price of a pair of shoes was £124. Work out the sale price for the shoes.

$\frac{35}{100} = 0.35$ and $1 - 0.35 = 0.65$

> Work out the multiplier by subtracting the multiplier for 35% from 1.

Sale price = 124 × 0.65 = £80.60

WORKIT!

Jarinda is buying a computer from a shop that offers interest-free credit.

The computer costs £450. She has to pay a deposit of 20% and then a fixed monthly payment over 2 years. Work out how much the fixed monthly payment will be, without using a calculator.

10% of £450 = £45

> Divide by 10 to find 10%.

20% of £450 = £90

> Double 10% to find 20%.

Deposit = £90

Remainder to be paid over 2 years = 450 − 90 = £360

Monthly payment = $\frac{360}{24}$ = £15

NAILIT!

Use the easiest method that works. Here, you don't need to work out 1%, only 10%, as 20% is a simple multiple of that.

Finding the original value

Sometimes you are given a value after a certain percentage is added and need to find the original value. For example, most shops show a price with the 20% VAT included so you need to calculate the original price before the VAT was added.

WORKIT!

1 A car costs £18 000 including VAT at 20%. Find the cost of the car before the VAT was added.

120% of original price = 18 000

10% of original price = $\frac{18\ 000}{12}$ = 1500

100% of original price = 1500 × 10 = 15 000

Original price = £15 000

> The price after the discount is (100 − 15) = 85% of the original price.

2 A tablet computer costs £204 after a discount of 15% in a sale. Calculate the original price of the computer before the discount.

85% of original price = 204

5% of original price = $\frac{204}{17}$ = 12

100% of original price = 12 × 20 = 240

Original price = £240

> You can spot that 5 × 17 = 85.

> You cannot get the correct answer by finding 20% of £18000 and then subtracting it from £18000. This is because the 20% is of the original price, which is a smaller amount.

> The price before the VAT was 100%, so the price including VAT is 120% of the original price.

> Alternatively, 18 000 ÷ 1.2 = 15 000.

NAILIT!

When you read the question, decide whether your answer should be larger or smaller than the original value – then check it is when you've finished.

DOIT!

Write down how to find the multiplier for percentage increase and decrease questions.

Simple interest

If you put money into a savings account, you will be paid a percentage of the amount in interest. Simple interest means that the amount paid is not added to the original amount and re-invested (for example because it is withdrawn) so the original amount stays the same.

To work out simple interest:

1 Work out the interest that would be paid at the end of one year.

2 Multiply this amount by the number of years the money is invested to give the total interest paid.

WORKIT!

£10 000 is invested in an account paying simple interest of 4.5% each year. Find the total amount of interest earned over 5 years.

Amount of interest in one year = 4.5% of £10 000 = $\frac{4.5}{100}$ × 10 000 = £450

Total interest paid over 5 years = 5 × £450 = £2250

CHECKIT!

1 In a batch of 300 eggs, 8 eggs were bad. Work out the number of bad eggs as a percentage of the total number of eggs. Give your answer to 2 decimal places.

2 A football manager earns £600 000 per year. On the promotion of his team to the Premier League, his earnings increase to £1.1 million per year. Find the percentage increase in his pay. Give your answer to 1 decimal place.

3 Jasmeen earns a salary of £38 000 per year in her job as a nurse. She is awarded a pay rise of 3.5%. Find her new salary.

4 In a sale, a mobile phone is reduced by 18%. The sale price is £291.92. Work out the original price.

5 £12 000 is invested at an interest rate of 3.5% for 6 years. If simple interest is paid, find the total amount of interest paid over the 6 years.

Direct and inverse proportion

Direct proportion

$y \propto x$ means that y is **directly proportional** to x: as x increases so does y.
For example, doubling x will double y, and halving x with halve y.

The proportional sign can be replaced by an equals sign provided a constant, usually called k is included. k is called the **constant of proportionality** and can be found by substituting known values into the equation for x and y.

For direct proportion, $y = kx$.

If y is directly proportional to x^2 we write $y = kx^2$.

WORKIT!

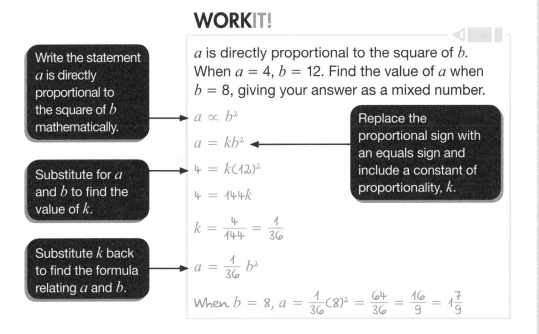

Write the statement a is directly proportional to the square of b mathematically.

Substitute for a and b to find the value of k.

Substitute k back to find the formula relating a and b.

a is directly proportional to the square of b.
When $a = 4$, $b = 12$. Find the value of a when $b = 8$, giving your answer as a mixed number.

$a \propto b^2$

$a = kb^2$

Replace the proportional sign with an equals sign and include a constant of proportionality, k.

$4 = k(12)^2$

$4 = 144k$

$k = \dfrac{4}{144} = \dfrac{1}{36}$

$a = \dfrac{1}{36} b^2$

When $b = 8$, $a = \dfrac{1}{36}(8)^2 = \dfrac{64}{36} = \dfrac{16}{9} = 1\dfrac{7}{9}$

NAILIT!

Always check back to see how the question asks the answer to be given, in this case a mixed number.

Currency problems

Converting between currencies is a particular example of direct proportion. An amount of money in one currency is proportional to the amount in another currency.

WORKIT!

Alegra is on holiday in the USA and sees a watch for $120.

The exchange rate is £1 = $1.45.

a Work out the cost of the watch in pounds.

Cost of watch in pounds $= \dfrac{120}{1.45} = £82.76$

Amount in $ = 1.45 × amount in £.

Commission is a charge that is a percentage of the value of the amount of money changed.

b When Alegra returns to the UK she has $300 left over which she wants to change back into pounds.

The travel agent offers an exchange rate of £1 = $1.32 without commission. The bank offers an exchange rate of £1 = $1.28 with 3% commission.

Should she use the travel agent or the bank to convert her money back into pounds? Explain your answer.

Travel agent: Amount in pounds $= \frac{300}{1.32} = £227.27$

Bank: Amount in pounds $= \frac{300}{1.28} = £234.38$

The 3% commission is worked out and then deducted.

Commission $= \frac{3}{100} \times 234.38 = £7.03$

Net amount from bank $= £234.38 - £7.03 = £227.35$

She should use the bank.

Inverse proportion

WORKIT!

It takes 3 men 5 hours to put up a fence. How long would it take 6 men?

As the number of men doubles, the number of hours halves.

$5 \div 2 = 2.5$ hours

In the previous example, the number of men is **inversely proportional** to the time it takes. If y is **inversely proportional** to x, it can be written as $y \propto \frac{1}{x}$. This means if x doubles, the value of y halves. Inverse proportion can be written with a constant of proportionality, k, as $y = \frac{k}{x}$.

As for direct proportion, if you know a pair of values for x and y, then you can find the value of k.

WORKIT!

Here are some statements. For each statement, write down an equation including a constant.

a a is directly proportional to b. $a = kb$

b R is inversely proportional to I.   Also, $I = \frac{k}{R}$ ➤ $R = \frac{k}{I}$

c y is directly proportional to the square of x. $y = kx^2$

d The time of a swing of a pendulum, T, in seconds is directly proportional to the square root of its length, l, $T = k\sqrt{l}$
in metres.

e a is inversely proportional to the square root of b. $a = \frac{k}{\sqrt{b}}$

WORKIT!

The pressure, P, of a gas is inversely proportional to its volume, V.
When $V = 12$, $P = 5$.

a Find a formula for P in terms of V.

$P \propto \frac{1}{V}$ so $P = \frac{k}{V}$

$5 = \frac{k}{12}$, giving $k = 60$ ◄———

> Substitute $P = 5$ and $V = 12$ to find the value of k.

$P = \frac{60}{V}$

b Calculate the value of P when $V = 6$.

$P = \frac{60}{V}$

$= \frac{60}{6}$

$= 10$

DO IT!

Design and annotate a revision card for finding the constant of proportionality.

✓ CHECKIT!

1 L is inversely proportional to n.

Carlos says that if you double n, then L will double.

Explain why Carlos is wrong.

2 y is directly proportional to x.

  a Write an equation showing the relationship between x and y.

  b When $y = 8$, $x = 3$. Find the value of y when $x = 4$. Give your answer to 1 d.p.

3 Rosie is on holiday in Greece. She sees a pair of sunglasses costing €120.

She looks on the internet and finds that the cheapest price in the UK is £89.

The exchange rate is £1 = €1.27.

  a In which country are the sunglasses cheaper?

  b Work out the difference in the cost of the sunglasses in the UK and Greece. Give your answer in £.

4 The volume, V cm³, of a sphere is directly proportional to the cube of its radius, r cm.

A sphere with a radius of 2 cm has a volume of 33.5 cm³.

Find the volume of a sphere with a radius of 4 cm.

5 The pressure of a gas, P, is inversely proportional to the volume it occupies, V.

When $V = 1$, $P = 100\,000$. Work out P when $V = 3$. Give your answer to the nearest whole number.

6 $a \propto b^2$ and $a = 96$ when $b = 4$. Find a when $b = 5$.

7 The surface area of a cube, A cm², is directly proportional to the square of the side of the cube, x cm.

  a Write the relationship as an equation and find the value of the constant of proportionality.

  b Hence write down the surface area of a cube with side of length 4 cm.

Graphs of direct and inverse proportion and rates of change

Direct proportion

A graph showing direct proportion is a straight line passing through the origin.

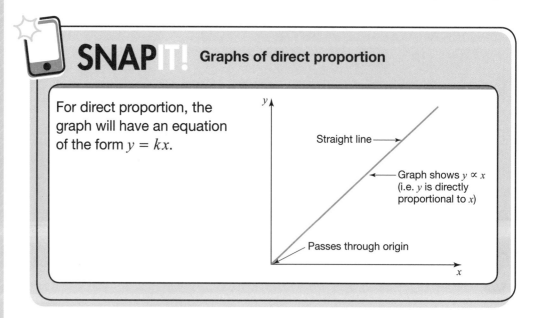

SNAP IT!   Graphs of direct proportion

For direct proportion, the graph will have an equation of the form $y = kx$.

Straight line

Graph shows $y \propto x$ (i.e. y is directly proportional to x)

Passes through origin

This equation is of the form $y = mx + c$, a straight line with gradient m and intercept on the y-axis c. The line passes through the origin, so $c = 0$. So the value of k will be the gradient of the line.

Inverse proportion

The graph showing inverse proportion is a curve of the type shown below.

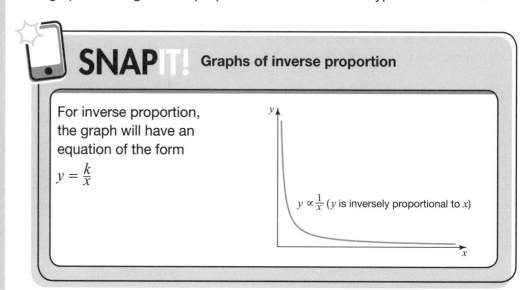

SNAP IT!   Graphs of inverse proportion

For inverse proportion, the graph will have an equation of the form

$y = \dfrac{k}{x}$

$y \propto \dfrac{1}{x}$ (y is inversely proportional to x)

WORKIT!

A graph is drawn of a quantity V against a quantity c.

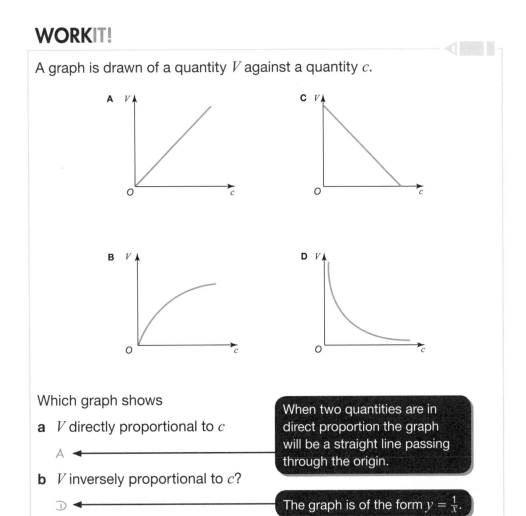

Which graph shows

a V directly proportional to c

A

> When two quantities are in direct proportion the graph will be a straight line passing through the origin.

b V inversely proportional to c?

D

> The graph is of the form $y = \frac{1}{x}$.

Calculating the rate of change from a graph

Graphs used to determine the **rate of change** of a quantity plotted on the y-axis always have **time** along the x-axis.

Suppose an experiment is conducted to investigate the rate of production of oxygen gas during a chemical reaction. The volume of oxygen in m³ is on the y-axis and the time in seconds is on the x-axis. The rate of change of the volume of oxygen is found by measuring the gradient of the graph.

WORKIT!

The volume of oxygen (O_2) produced in a chemical reaction was measured at certain times and gave the following results.

Time (s)	0	50	100	150	200	250	300
Volume O_2 (m³)	0	5.0	10.0	14.8	19.0	22.5	25.0

Plot these results on a graph with volume of oxygen on the y-axis and time on the x-axis.

Determine the initial rate of production of oxygen in m^3/s.

Make sure that you label both axes and include the units.

The points are almost in a straight line initially but then they start to bend.

Draw a straight line up to about 200 s and then a smooth curve after that. As the start of the graph is a straight line we can say that the gradient and hence the rate of change of volume of O_2 is constant.

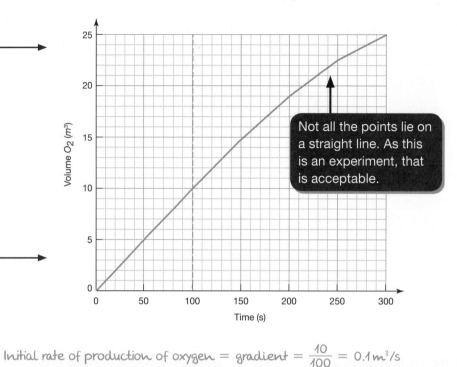

Not all the points lie on a straight line. As this is an experiment, that is acceptable.

Initial rate of production of oxygen = gradient = $\frac{10}{100}$ = 0.1 m^3/s

To find the gradient, draw a triangle in the straight section of the graph.

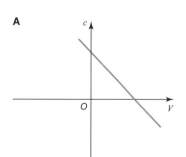 **DOIT!**

Draw a poster of annotated graphs showing direct and inverse proportion.

CHECKIT!

1 Two quantities, c and V, are in inverse proportion. Which graph shows this?

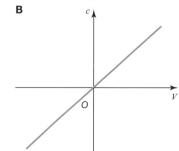
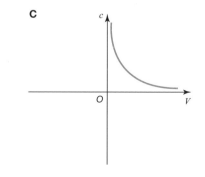

A   **B**   **C**

2 Quantity a is directly proportional to quantity b. Which one of the following graphs shows this?

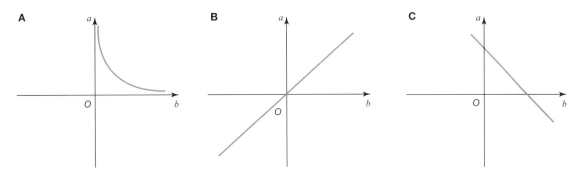

A   B   C

3 The graph shows two quantities, P and V, that are inversely proportional to each other.

The points $A(3, 12)$ and $B(6, a)$ lie on the curve.

Find the value of a.

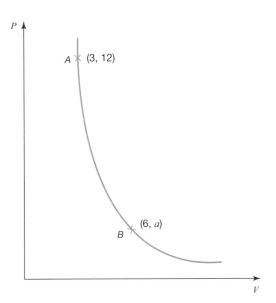

4 The graph shows two quantities x and y which are in inverse proportion to each other.

Find the values of coordinates a and b.

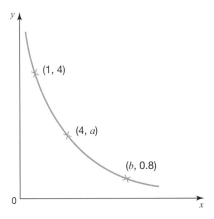

5 For the values below, y is directly proportional to x^2. If a is a positive number, work out the value of a.

x	2	a
y	16	36

Growth and decay

Compound interest

With **compound interest**, the interest is added onto the total amount at the end of each year, which then earns interest too. This means the overall increase is greater each year.

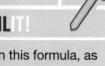
SNAPIT!   Compound interest

Amount at the end of n years $= A_0 \times (\text{multiplier})^n$
where n = time period in years, A_0 = original amount of money, and multiplier $= 1 + \frac{\% \text{ interest rate}}{100}$.

WORKIT!

Jack invests £2000 at 2.75% compound interest. Find the value of the investment at the end of 3 years.

Multiplier $= 1 + \frac{2.75}{100} = 1.0275$ ← Work out the multiplier.

Amount at the end of 3 years $= 2000 \times (1.0275)^3 = £2169.58$

Use the formula with $A_0 = 2000$ and $n = 3$.

As this is a money question, give your answer to the nearest penny.

Compound growth

Compound interest is one example of **compound growth**. Other examples can be found in biology. The formula is the same, although the periods of time may be days or hours, not years.

SNAPIT!   Compound growth

Amount at the end of n periods of time $= A_0 \times (\text{multiplier})^n$
where n = number of time periods (e.g. years), A_0 = original amount, and multiplier $= 1 + \frac{\% \text{ rate of growth}}{100}$.

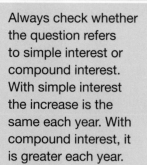

DOIT!

Calculate the value of a £1000 investment after 3 years at an interest rate of 5%, for both simple interest and compound interest. Produce a poster showing your results.

WORKIT!

1 A colony of bacteria increases by 12% each day. There are 4000 bacteria at the start. How many bacteria will there be after 3 days? Give your answer to the nearest whole number.

Multiplier $= 1 + \frac{12}{100} = 1.12$

Amount at the end of 3 days $= 4000 \times 1.12^3 = 5620$ (to 4 s.f.)

2 Algae in a pond grow very quickly in warm weather. The number of algae is 200 000 after 5 days. If the algae grow at a rate of 23% per day, how many algae were there originally?

$\text{Multiplier} = 1 + \frac{23}{100} = 1.23$

Amount at the end of n days
$= A_0 \times (\text{multiplier})^n$

$200\,000 = A_0 \times (1.23)^5$

$\frac{200\,000}{(1.23)^5} = A_0$

$A_0 = 71\,040$

> **Substitute the amount after 5 days and $n = 5$.**

SNAPIT!

Compound decay

Amount at the end of n years
$= A_0 \times (\text{multiplier})^n$

where n = time period in years, A_0 = original amount and

$\text{multiplier} = 1 - \frac{\% \text{ rate of decay}}{100}$.

Compound decay

Compound decay is where a quantity goes down by a certain percentage – like compound growth but in reverse. For example, when you buy a new car it depreciates (i.e. the value goes down with time).

For compound decay, the formula is the same as for compound growth but the multiplier is less than 1.

> **For decay questions, the multiplier is calculated by subtracting the rate from 1.**

WORKIT!

Millie buys a new car costing £18 500. The car depreciates at an average rate of 15% per year. Find the value of the car after 4 years, giving your answer to the nearest whole number.

$\text{Multiplier} = 1 - \frac{15}{100} = 0.85$

Amount at the end of 4 years $= 18\,500 \times (0.85)^4 = 9657.12 = £9657$
(to nearest whole number).

CHECKIT!

1 The equation for compound growth and decay is:

Amount at the end of n units of time
$= A_0 \times (\text{multiplier})^n$.

Find the multiplier for each of these situations.

a pond weed in a pond where the growth rate is 5% per week

b bacterial growth where growth rate is 25% per hour

c increase in money if the rate of compound interest is 3.75% per year

d decrease in value if the rate of depreciation is 21%.

2 A motorbike depreciates at a rate of 18% per year. Ben buys a new motorbike costing £9000. How much will his motorbike be worth after 3 years? Give your answer to the nearest whole number.

3 A restaurant chain currently has 4000 restaurants across the UK. It has been expanding by 25% each year. How many restaurants did the company have 3 years ago?

Ratios of lengths, areas and volumes

Scale factors

When two shapes are **similar**, they are identical in shape but not in size. One shape is simply an **enlargement** of the other. A missing length on one of the shapes can be found by considering the scale factor:

$$\text{Scale factor for enlargement} = \frac{\text{big}}{\text{small}}$$

$$\text{Scale factor for reduction} = \frac{\text{small}}{\text{big}}$$

WORKIT!

P and Q are similar triangles. Find the length of the side marked x on triangle Q.

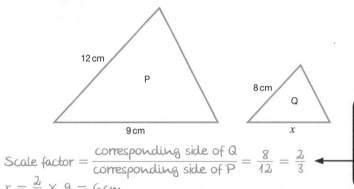

Multiply the side on P corresponding to x by the scale factor.

You are finding the smaller length, so find the scale factor for the reduction.

$$\text{Scale factor} = \frac{\text{corresponding side of Q}}{\text{corresponding side of P}} = \frac{8}{12} = \frac{2}{3}$$

$$x = \frac{2}{3} \times 9 = 6\,cm$$

You can prove two triangles are similar by comparing the ratios of corresponding sides.

WORKIT!

Identify whether these triangles are similar.

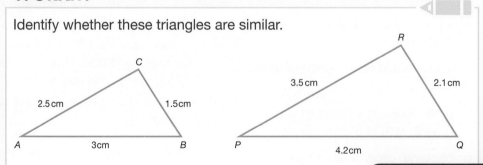

Work out the ratio for each corresponding pair of sides.

$$\frac{PQ}{AB} = \frac{4.2}{3} = 1.4 \qquad \frac{QR}{BC} = \frac{2.1}{1.5} = 1.4 \qquad \frac{PR}{AC} = \frac{3.5}{2.5} = 1.4$$

The ratio is the same, so the triangles are similar.

Scale factors for area and volume

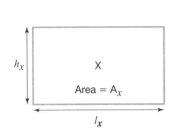
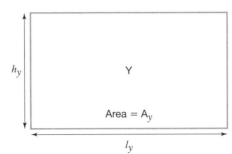

For rectangles X and Y, scale factor for length, $k = \dfrac{h_y}{h_x} = \dfrac{l_y}{l_x}$

Scale factor for area $= \dfrac{h_y \times l_y}{h_x \times l_x}$

$= \dfrac{h_y}{h_x} \times \dfrac{l_y}{l_x} = k^2$ or the square of the scale factor for length.

Similarly, the scale factor for volume is equal to the cube of the scale factor for length.

SNAP IT!

2D and 3D scale factors

For two similar shapes or solids with a scale factor k:

- scale factor for area $= k^2$
- scale factor for volume $= k^3$

WORKIT!

Prove that triangle ABC is mathematically similar to triangle DBE.

$AD = 3\,\text{cm}$, $DB = 6\,\text{cm}$ and $AC = 8\,\text{cm}$.

Work out the area of triangle DBE.

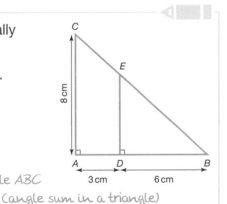

Angle CAB = angle EDB = 90°

Angle ABC = angle DBE (same angle)

Angle ACB = 180° − angle CAB − angle ABC
(angle sum in a triangle)

Angle DEB = 180° − angle EDB − angle DBE (angle sum in a triangle)

So angle ACB = angle DEB

All the angles are the same, so triangles ABC and DBE are similar. ◀— First prove that the triangles are similar.

Area of triangle $ABC = \frac{1}{2} \times b \times h = \frac{1}{2} \times 9 \times 8 = 36\,\text{cm}^2$

Scale factor $= \dfrac{DB}{AB} = \dfrac{6}{9} = \dfrac{2}{3}$ ◀— Work out the scale factor from the ratio of corresponding sides.

$\dfrac{\text{Area } DBE}{\text{Area } ABC} = (\text{scale factor})^2 = \left(\dfrac{2}{3}\right)^2 = \dfrac{4}{9}$

Area $DBE = \frac{4}{9} \times$ Area $ABC = \frac{4}{9} \times \frac{1}{2} \times 9 \times 8 = 16\,\text{cm}^2$ ◀— Use the fact that the triangles are similar to write the ratio of the areas.

WORKIT!

Two solid spheres, P and Q, are mathematically similar.
The ratio of the volume of P to the volume of Q is 8 : 125.

a Find the ratio of the radius of sphere P to the radius of sphere Q.

$$\frac{V_P}{V_Q} = \left(\frac{r_P}{r_Q}\right)^3$$

Scale factor for volume = (scale factor for length)³.

$$\frac{8}{125} = \left(\frac{r_P}{r_Q}\right)^3$$

$$\frac{r_P}{r_Q} = \sqrt[3]{\frac{8}{125}} = \frac{2}{5}$$

b The surface area of sphere Q is 37.5 cm². Show that the surface area of sphere P is 6 cm².

Scale factor for area = (scale factor for length)².

$$\frac{A_P}{A_Q} = \left(\frac{r_P}{r_Q}\right)^2 = \left(\frac{2}{5}\right)^2 = \frac{4}{25}$$

$$A_P = \frac{4}{25} \times A_Q = \frac{4}{25} \times 37.5 = \frac{150}{25} = 6$$

Surface area of sphere P is 6 cm².

DOIT!

Make up some scale factor questions then come back to them tomorrow.

✓ CHECKIT!

1 Triangles A and B are similar. Find the length of the side marked x on triangle A. Give your answer as a mixed number.

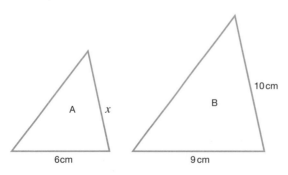

6cm   9 cm   10 cm

A   x   B

2 Two cylinders are mathematically similar. Cylinder A has a volume of 27 cm³ and cylinder B has a volume of 64 cm³.

The surface area of cylinder B is 96 cm².

Show that the surface area of cylinder A is 54 cm².

3

5 cm   5 cm

A   B   C

E

8 cm

D

ABC and AED are straight lines. Lines BE and CD are parallel. Angle ACD = 90°.

CD = 8 cm, BC = 5 cm and AB = 5 cm.

a Work out the length of side BE.

b Work out the area of triangle ABE.

Gradient of a curve and rate of change

The gradient of a curve changes with position so you need to know where the gradient is to be measured. Usually the x coordinate is given, which determines where on the curve to find the gradient.

With rate of change graphs, time is always plotted on the x-axis.

The instantaneous rate of change

The **instantaneous rate of change** is the gradient of the tangent drawn at that point.

For example, this graph shows how the velocity of an object changes with time.

The gradient of a velocity–time graph gives the acceleration. To find the acceleration at time $t = 10\,\text{s}$, work out the gradient of the tangent:

Acceleration at 10 s = gradient
$= \dfrac{12 - 4}{16 - 6} = \dfrac{8}{10} = 0.8\,\text{m/s}^2$

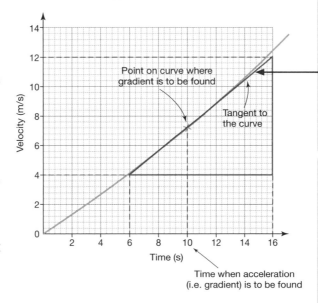

Point on curve where gradient is to be found

Tangent to the curve

Time when acceleration (i.e. gradient) is to be found

The tangent is the straight line that just touches the curve.

STRETCH IT!

Use the same method to find the gradient of a curve from a graph by drawing in a tangent.

The average rate of change

The **average rate of change** is the gradient of the line drawn between two points on the curve. Each point corresponds to a certain time.

Average acceleration between 6 s and 16 s = gradient = $\dfrac{8}{10}$
$= 0.8\,\text{m/s}^2$

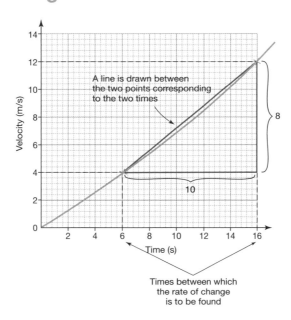

A line is drawn between the two points corresponding to the two times

Times between which the rate of change is to be found

WORKIT!

The graph shows how the height of grain in a large hopper changes with time, t, when it is being filled with grain.

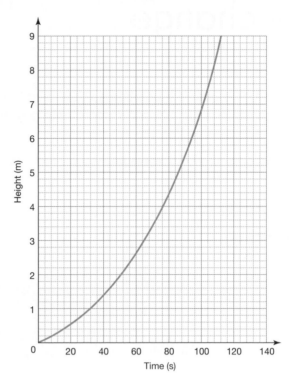

NAILIT!

The axes are time in seconds and height in metres. You could also have labelled them t and $f(t)$. Note that s and m are units not variables so these could not be used to label the axes on their own.

a Find the instantaneous rate of change of height when $t = 80\,\text{s}$.

Draw a tangent to the curve at $t = 80\,\text{s}$ and find its gradient.

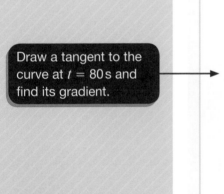

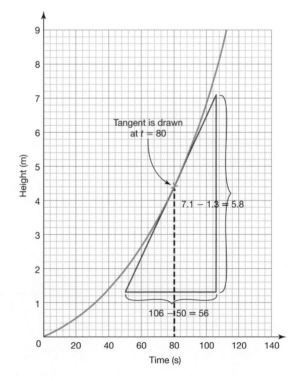

Tangent is drawn at $t = 80$

$7.1 - 1.3 = 5.8$

$106 - 50 = 56$

Instantaneous rate of change of height $=$ gradient $= \dfrac{5.8}{56}$

$\qquad\qquad\qquad = 0.1\,\text{m/s (to 1 s.f.)}$

b Find the average rate of change of height between $t = 40\,\text{s}$ and $t = 80\,\text{s}$.

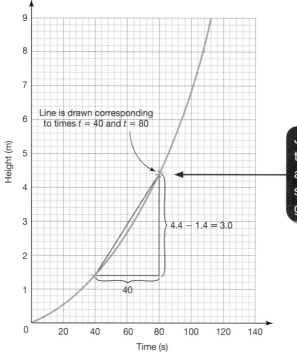

Line is drawn corresponding to times $t = 40$ and $t = 80$

Join the points on the curve at $t = 40\,\text{s}$ and $t = 80\,\text{s}$ with a straight line and find its gradient.

$4.4 - 1.4 = 3.0$

40

Average rate of change of height between $t = 40\,\text{s}$ and $t = 80\,\text{s}$

$= \text{gradient} = \dfrac{3}{40} = 0.08\,\text{m/s}$ (to 1 s.f.)

DOIT!

Draw and annotate distance–time and velocity–time graphs to show instantaneous and average gradients and what they represent.

CHECK**IT!**

1 An accelerating particle has the velocity–time graph shown on the right.

a State what the gradient of the curve represents.

b Find the average acceleration of the particle between 2 s and 6 s.

c Find the time when the instantaneous acceleration is the same as the average acceleration calculated in part b.

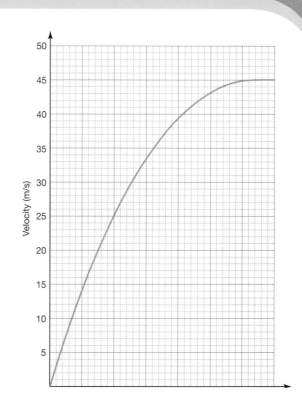

Converting units of areas and volumes, and compound units

Converting areas and volumes

Converting between m² and cm²

A 1 m by 1 m square has an area of 1 m².

This is the same as a 100 cm by 100 cm square, which has an area of 100 × 100 = 10000 cm².

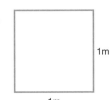
1m

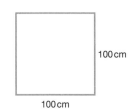
100cm

> **Note that 1 m² does not equal 100 cm².**

> **We've given one answer in standard form and one as a decimal. Both are acceptable.**

So $1\,m^2 = 10\,000\,cm^2$ and $1\,cm^2 = \frac{1}{10\,000}\,m^2 = 1 \times 10^{-4}\,m^2$.

Similarly, $1\,cm^2 = 100\,mm^2$ and $1\,mm^2 = \frac{1}{100}\,cm^2 = 0.01\,cm^2$.

Converting between m³ and cm³

Similarly, a cube of side 1 m has a volume of 1 m³, and a cube of side 100 cm has a volume of 100 × 100 × 100 = 1 000 000 cm³.

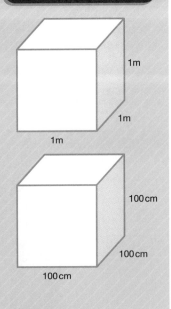

So $1\,m^3 = 1\,000\,000\,cm^3$ and $1\,cm^3 = \frac{1}{1\,000\,000}\,m^3 = 1 \times 10^{-6}\,m^3$.

Similarly, $1\,cm^3 = 1000\,mm^3$ and $1\,mm^3 = \frac{1}{1000}\,cm^3 = 0.001\,cm^3$.

WORKIT!

1 Convert:

  a 250 cm³ to m³

$$250\,cm^3 = 250 \times 10^{-6}\,m^3 = 2.5 \times 10^{-4}\,m^3$$

  b 1.500 m² to cm²

$$1.5\,m^2 = 1.5 \times 10\,000\,cm^2 = 1.5 \times 10^4\,cm^2$$

Compound units

> **The per symbol (i.e. /) acts like a divide sign, and can tell you information about the formula. For example, one unit of speed is km/h, which tells you a distance in km is divided by a time in h.**

Compound units are made up of two units (for example m/s, km/h, g/cm³). Quantities such as speed, density, pressure, rate of pay and price per unit have compound units. These are all worked out by dividing one quantity with one unit by a different quantity with a different unit.

SNAP IT! Compound units

Compound units include:

$$\text{speed} = \frac{\text{distance}}{\text{time}} \qquad \text{unit m/s or km/h}$$

$$\text{density} = \frac{\text{mass}}{\text{volume}} \qquad \text{unit g/cm}^3 \text{ or kg/m}^3$$

$$\text{pressure} = \frac{\text{force}}{\text{area}} \qquad \text{unit N/m}^2 \text{ or Pa } (= \text{N/m}^2)$$

$$\text{rate of pay} = \frac{\text{pay}}{\text{time}} \qquad \text{unit £/h, £/month, £/year}$$

$$\text{price per unit} = \frac{\text{price}}{\text{number of units}} \qquad \text{unit for price on the top could be pence or pounds}$$

unit on the bottom could be g, kg, ml, tonne, etc.
compound unit p/g, £/kg, etc.

Speed is worked out using the formula

You need to be able to rearrange this

formula to get: distance = speed × time

$$\text{speed} = \frac{\text{distance}}{\text{time}}$$

$$\text{time} = \frac{\text{distance}}{\text{speed}}$$

STRETCH IT!

When speed is
in a particular
direction, it is
called **velocity**.
For more on this
see pages 121–3.

NAIL IT!

You need to learn this formula as it will not be given. To remember the
formula for speed, think of the units for speed, which are m/s or km/h:
distance divided by time.

WORKIT!

1 Dhaya is driving her car at an average speed of 80 km/h.

a Work out her speed in metres per second, correct to the nearest
whole number.

$$80\,\text{km/h} = 80 \times 1000\,\text{m/h}$$

$$= \frac{80 \times 1000}{60 \times 60}\,\text{m/s}$$

$$= 22.22\,\text{m/s}$$

$$= 22\,\text{m/s (to nearest whole number)}$$

b Calculate the time in minutes it would
take her to drive a distance of 30 km at
this average speed.

Rearrange speed $= \frac{\text{distance}}{\text{time}}$
to give time $= \frac{\text{distance}}{\text{speed}}$

$$\text{time} = \frac{\text{distance}}{\text{speed}} = \frac{30}{80} = 0.375\,\text{h}$$

$$0.375\,\text{h} = 0.375 \times 60\,\text{minutes} = 22.5\,\text{minutes}$$

NAIL IT!

Always look back at
the question to check
how accurate the
answer needs to be.

NAIL IT!

Always look at the
units required for the
answer. This tells you
whether you need
to change any of the
units for the values in
the calculation.

The answer needs to be in N/m², so convert the area from cm² to m².

NAILIT!

Remember that time is not metric.

2 Using the formula pressure = $\frac{\text{force}}{\text{area}}$ work out the pressure in N/m² when a force of 500 N acts on an area of 20 cm².

$$20\,cm^2 = 20 \times 10^{-4}\,m^2 = 2 \times 10^{-3}\,m^2$$

$$\text{pressure} = \frac{\text{force}}{\text{area}} = \frac{500}{2 \times 10^{-3}} = 250\,000\,N/m^2$$

3 A train covers a distance of 29 km in 45 minutes. Calculate its average speed in km/h. Give your answer to the nearest whole number.

$$45\,mins = \frac{45}{60}\,h = 0.75h$$

$$\text{speed} = \frac{\text{distance}}{\text{time}} = \frac{29}{0.75} = 38.67\,km/h = 39\,km/h \text{ (to nearest hour)}$$

DO IT!

Look up the 'speed formula triangle' on the internet. Produce similar triangles for density and pressure.

CHECK IT!

1 The diagram shows a solid block of wood.

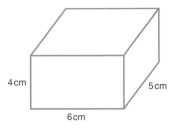

4 cm   5 cm
6 cm

a Find the total surface area of the block in:

  i mm²   **ii** m².

b Find the volume of the block in m³.

2 A bar of gold has a volume of 600 cm³ and a mass of 1.159 kg. Work out the density of gold in g/cm³ giving your answer correct to 3 decimal places.

3 A block of steel with a volume of 1 m³ is melted and used to create lots of ball bearings. The volume of each ball bearing is 0.5 cm³. How many ball bearings could be made?

4 A car is travelling with an average speed of 50 km/h. Find this speed in m/s. Give your answer correct to 2 decimal places.

5 Kieran drove from Manchester to Bristol. It took him 4 hours at an average speed of 70 km/h.

Mary drove by the same route and took 5 hours.

a What was Mary's average speed for her journey?

b If Mary took a different route, how might this affect your answer to part a?

1 In a sale the price of a dress was reduced by 20% to £76.80. What was the original price of the dress?

2 The area of floating pond weed that grows on the surface of a pond in t days is $A\,\text{m}^2$, where

$A_0 = 5$

$A_{t+1} = 1.02A_t$

Work out the area covered by the pond plants in

a 1 day   **b** 3 days.

Give your answers to 2 decimal places.

3 y is inversely proportional to x.

When $x = 2.5$, $y = 4$.

Find the value of y when $x = 5$.

4 The ratio of the number of students studying history in Year 11 to those not studying history is $2:7$. There are 54 students studying history. How many students are there in Year 11?

5 The price of a house in a certain town has increased by 120% over the last 10 years.

The original price of the house was £220 000. How much is the house worth now? Give your answer to the nearest £5000.

6 £2000 is invested at an interest rate of 2.5% for 5 years. Work out how much money is in the account after 5 years if the interest paid is

a simple interest

b compound interest (to the nearest penny).

7 Tom goes on holiday to China. The exchange rate is £1 = 8.55 yuan.

He changes £400 into yuan.

a Work out how many yuan he should get.

After his holiday, Tom has 800 yuan left over which he wants to change back into pounds.

The travel agent offers an exchange rate of £1 = 8.6 yuan with 2.5% commission.

The post office offers an exchange rate of £1 = 8.9 yuan with no commission.

b Should Tom change his money at the travel agent or the post office? Explain your answer.

8 A company is owned by Joshua, Amy and Luke, in the ratio of $3:5:7$.

Any profits are shared in this ratio.

When the profits for a certain year were shared, Luke received £4000 more than Amy.

Work out the total profit for the year.

9 Two solids are mathematically similar. Solid A has a surface area of $25\,\text{cm}^2$ and solid B has a surface area of $4\,\text{cm}^2$.

The volume of solid A is $10\,\text{cm}^3$.

Find the volume of solid B.

10 The number of male and female members of a swimming club is in the ratio $7:4$.

25% of the male members are junior members.

10% of the female members are junior members.

Work out the percentage of junior members in the club, giving your answer to the nearest integer.

11 A bronze alloy is made by mixing copper and tin. The masses of the copper and tin are in the ratio $9:1$.

The density of copper is $8.9\,\text{g/cm}^3$.

The density of tin is $7.3\,\text{g/cm}^3$.

Work out the density of the bronze alloy. Give your answer to 1 decimal place.

Geometry and measures

2D shapes

You need to be familiar with the following terms.

Polygon: a figure with straight edges.

Irregular polygon: a polygon where the sides and angles are not all equal. For example, the diagram shows an irregular hexagon.

Regular polygon: a polygon where all the sides and angles are equal.

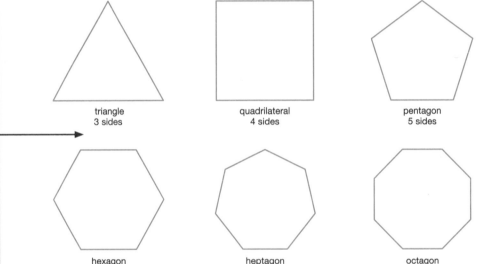

triangle
3 sides

quadrilateral
4 sides

pentagon
5 sides

hexagon
6 sides

heptagon
7 sides

octagon
8 sides

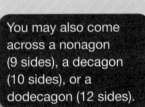

You may also come across a nonagon (9 sides), a decagon (10 sides), or a dodecagon (12 sides).

For regular polygons, the order of rotational symmetry and the number of lines of symmetry are both the same as the number of sides.

Order of rotational symmetry: the number of times you can rotate the shape onto itself so it fits exactly in one complete revolution.

Lines of symmetry: the number of mirror lines.

SNAP IT! Types of triangle

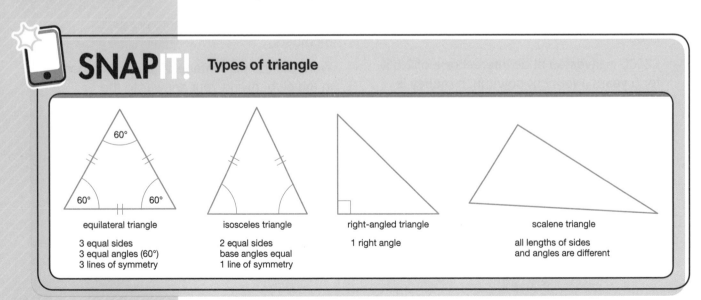

equilateral triangle

3 equal sides
3 equal angles (60°)
3 lines of symmetry

isosceles triangle

2 equal sides
base angles equal
1 line of symmetry

right-angled triangle

1 right angle

scalene triangle

all lengths of sides
and angles are different

SNAP IT!   Types of quadrilateral

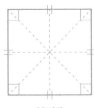

square

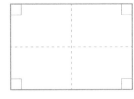

rectangle

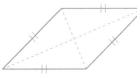

rhombus

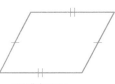

parallelogram

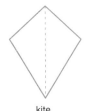

kite

trapezium   trapezium

Name	Pairs of parallel sides	Lines of symmetry	Order of rotational symmetry	Diagonals same length?	Equal sides	Equal angles
Square	2	4	4	Yes	Yes	All 90°
Rectangle	2	2	2	Yes	2 pairs	All 90°
Rhombus	2	2	2	No	Yes	2 pairs
Parallelogram	2	0	2	No	2 pairs	2 pairs
Kite	0	1	1	No	2 pairs	1 pair
Trapezium	1	1 or 0	1	Can be	1 pair or none	2 pairs, 1 pair or none

With a rhombus, a square and a kite, the diagonals intersect at right angles. For a square they are the same length, but not for a rhombus or kite.

DO IT!

Draw a flowchart for finding out the type of triangle or quadrilateral from information about its sides, angles and symmetry.

CHECK IT!

1 Here are some statements about polygons. State whether each statement is true or false.

a A trapezium always has one pair of parallel sides.

b The diagonals of a rhombus are the same length.

c It is possible to have a right-angled isosceles triangle.

d A parallelogram has order of rotational symmetry of 1.

e The order of rotational symmetry of a regular polygon is the same as the number of sides.

f An equilateral triangle can only have internal angles of 60°.

Constructions and loci

A bisector divides something into two equal parts.

Constructions

Using only a ruler and compasses, you will be asked to **construct** angles, lines and figures. Here are the constructions you need to be able to draw.

Perpendicular bisector of a line segment

To bisect (cut in half) a line with another line at right angles:

1. Put the compass point on A and set the radius to about two-thirds of AB.

2. Without altering the radius, draw two arcs at approximate positions P and Q.

3. Repeat step 2, putting the compass point on B.

4. Draw the line between P and Q.

The line PQ is the set of points equal in distance from the points A and B.

Perpendicular to a line at a given point

To construct a line passing through point P and perpendicular to AB:

1. Put the compass point on P and set the radius to draw two arcs cutting AB.

2. Alter the radius so that it is about half the distance from point P to the line.

3. Using this radius, draw an arc from each of the intersections of the arc from step 1.

4. Draw a line from point P through the intersection of the arcs until it meets AB.

Perpendicular to a line through a given point on the line

The perpendicular from P is the shortest distance to the line AB.

To draw the perpendicular through a point on a line, use the same radius to draw two arcs on the line centred on P. Then use these two intersections as A and B and construct the perpendicular bisector of AB.

Bisecting an angle

The angle bisector is the set of points equal perpendicular distance from lines OA and OB.

To bisect (cut in half) angle AOB:

1. Put the compass point on O and draw two arcs cutting OA and OB.

2. Without altering the radius, draw two arcs from the intersections in step 1, to intersect at P.

3. Draw a line through P to O.

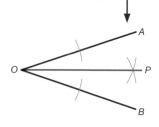

Loci

If the position of a point P moves in some way then all the possible positions point P can take is called the **locus** (plural **loci**) of P.

WORKIT!

AB is a straight line. Draw the locus of the point P that moves in such a way that it is always 2 cm from AB.

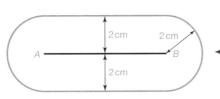

WORKIT!

Use only ruler and compasses to answer this question.

Point P lies inside the quadrilateral ABCD. It is the same distance from A and B and also a distance of 7 cm from point D. Mark on the diagram the position of P.

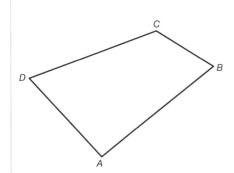
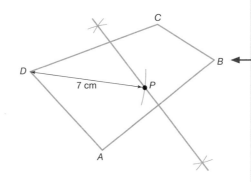

NAILIT!

A locus that stays the same distance from a point is either a circle or an arc of a circle.

Draw two semicircles of radius 2 cm with centres A and B. Join the tangents to these to give the locus.

Draw the perpendicular bisector of AB to give all the points equal distance from A and B.

Draw an arc centred on D with radius 7 cm to give all the points 7 cm from D.

P is where the line and arc intersect.

DOIT!

Write out a revision card for each construction.

CHECKIT!

1 OP and OQ are straight lines at an angle to each other. X travels in such a way that it is always the same distance between OP and OQ.

Copy the diagram and draw the locus of X.

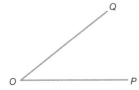

2 Two towns, A and B, are 70 km apart.

A wind farm is to be sited in an area around the two towns. It must be at least 30 km from town A and at least 60 km from town B.

Using a scale of 1 cm to 10 km, draw a map and shade the region where the wind farm could be situated.

Properties of angles

Angles at a point

Angles at a point on a straight line add up to 180°.
$x + y = 180°$

Vertically opposite angles are equal.

Angles at a point add up to 360°.
$a + b + c = 360°$

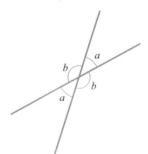

Properties of triangles

The angles in a triangle add up to 180°.
$a + b + c = 180°$

Isosceles triangle: equal base angles. The top angle can be found by adding the two base angles together and subtracting the result from 180°.

Equilateral triangle: all angles are 60° and all sides are equal.

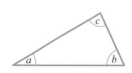

Parallel lines

Parallel lines are the same distance apart along their whole length.

Alternate angles are equal; on opposite sides of the line cutting the parallel lines; sometimes called 'Z' angles.

Corresponding angles are equal; sometimes called 'F' angles.

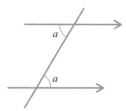

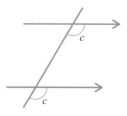

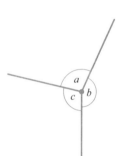

WORKIT!

AB and *CD* are two parallel lines.

a Work out, stating reasons, the size of angle *STR*.

> Angle *STP* is the same as angle *STR*.

angle STR = angle BRT = 70° (alternate angles)

b Work out, stating reasons, the size of angle *x*.

x = 180 − angle PST − angle STP (angle sum of a triangle)

= 180 − 50 − 70 = 60° (STP is the same angle as STR)

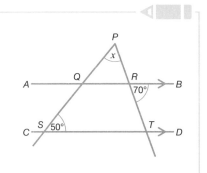

Proving that the angles in a triangle add up to 180°

You need to be able to prove that the angle sum of a triangle is 180°.

The angles marked *a* are alternate angles.

The angles marked *b* are alternate angles.

c is the top angle of the triangle.

$a + b + c = 180°$ (angles on a straight line)

> Draw two parallel lines as shown.

This is also the total of the angles in the triangle, so the angles in a triangle add up to 180°.

Angles in regular polygons

> Remember that regular polygons have equal sides and angles.

SNAP IT!   Angles in regular polygons

Exterior angle $= \dfrac{360°}{\text{number of sides}}$

Interior angle + exterior angle = 180°

Total of all the interior angles = number of sides × interior angle

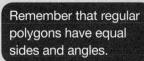

sides of a regular polygon

WORKIT!

The diagram shows a regular pentagon and a regular hexagon.

a Work out the size of the interior angle of a regular pentagon.

Exterior angle $= \dfrac{360°}{\text{number of sides}} = \dfrac{360}{5} = 72°$

Interior angle = 180 − 72 = 108°

> You need to spot that you need to find the interior angles of each of the polygons.

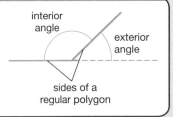

b Hence work out the size of angle *x*. Show all your working.

Exterior angle of regular hexagon $= \dfrac{360°}{\text{number of sides}} = \dfrac{360}{6} = 60°$

Interior angle = 180 − 60 = 120°

Angle x = 360 − (108 + 120) = 132°

> The angles at a point all add up to 360°.

DO IT!

Make two copies of the diagram. On one diagram, colour each pair of alternate angles a different colour. On the other, colour each pair of corresponding angles a different colour.

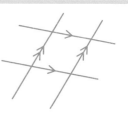

✓ CHECKIT!

1 Calculate the interior angle of a regular polygon with ten sides.

2 The exterior angle of a regular polygon is 40°.

  a Find the number of sides of the polygon.

  b Calculate the sum of all the interior angles of the polygon.

3 In the diagram, *AC* and *BE* are parallel. Find the size of angle *x*.

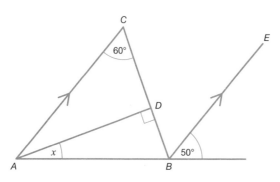

4 In the diagram, *AC*, *DF* and *GJ* are three parallel lines.

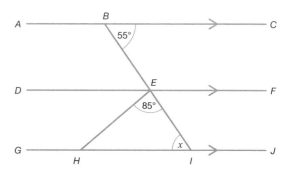

  a Write down, stating a reason, the size of angle *x*.

  b Write down, stating a reason, the size of angle *DEH*.

5 *ABCD* is a rhombus. Work out the angles of the rhombus.

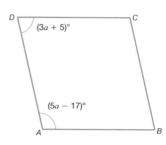

Congruent triangles

Two triangles are **congruent** if they fit exactly over each other. All the corresponding sides and corresponding angles are equal.

Triangles are congruent if at least three special facts are known.

SNAP IT!   Congruent triangles

The four possible **conditions for congruence** are that the following need to match:

- three sides (SSS)

- two angles and a corresponding side (ASA)

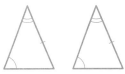

- two sides and the angle between them (SAS)

- for right-angled triangles only: right angle, hypotenuse and one other side (RHS)

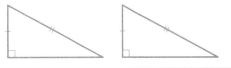

Once you have established the congruence using one of the above cases, you can use the fact that all the corresponding sides and angles are equal to find unknown angles and sides.

WORKIT!

Here are some pairs of triangles. Say if each pair is congruent or not. If they are congruent, give the reason (e.g. SAS).

a

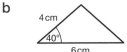

a Yes: RHS

b

b No

c

c Yes: SSS

d

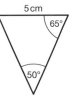

d Yes: ASA

e

e No

NAILIT!

If the triangles aren't the same way around, quickly draw them the same way around. This will help you to spot corresponding sides and angles.

WORKIT!

In the diagram, ABCD is a rhombus and AE = FC.

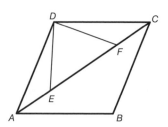

> Mark any equal sides and angles on a copy of the diagram.

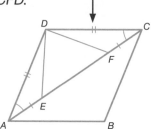

Prove that triangle AED is congruent to triangle CFD.

AE = FC (given in the question)

AD = CD (sides of a rhombus equal)

So triangle ACD is isosceles, so the base angles EAD and FCD are equal.

Triangle AED is congruent to triangle CFD (SAS).

DOIT!

Use congruent triangles to prove two of the constructions you covered on pages 130–2.

DOIT!

Create a mnemonic or a song to help you remember the conditions for congruence.

NAILIT!

In questions involving rhombuses and parallelograms, look for isosceles triangles, corresponding angles and alternate angles to prove congruency.

✓ CHECKIT!

1 In the quadrilateral, AB = AD and BC = CD. Prove that angle ABC = angle ADC.

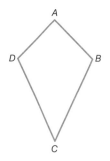

2 ABC is a triangle with AB = AC. M is the midpoint of BC.

Prove that angle AMB is a right angle.

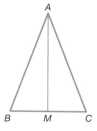

Transformations

When a transformation is applied to a shape, the original shape is called the **object** and the shape after the transformation is called the **image**.

Translations

With a **translation**, the shape moves according to a **vector** $\begin{pmatrix} x \\ y \end{pmatrix}$. Each x coordinate of the shape moves x units to the right if x is positive, or x units to the left if x is negative. Each y coordinate of the shape moves y units up if y is positive, or y units down if y is negative.

The diagram shows two different translations for the triangle P.

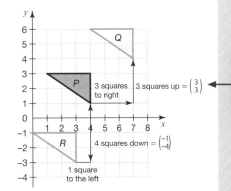

> Note that the shapes are identical under a translation (i.e. they are congruent).

Reflections

When a shape undergoes a **reflection** you need to know where the **mirror line** is.

Examples of mirror lines include:

- the y-axis (the vertical axis with the equation $x = 0$)

- the x-axis (the horizontal axis with the equation $y = 0$)

- $x = 4$ (a vertical line parallel to the y-axis at $x = 4$)

- $y = -2$ (a horizontal line parallel to the x-axis at $y = -2$)

- $y = x$ (a sloped line with a positive gradient, sloping at 45° if both axes have the same scale)

- $y = -x$ (a sloped line with a negative gradient, sloping at 45° if both axes have the same scale).

> Reflected shape after reflection in line $x = -2$.

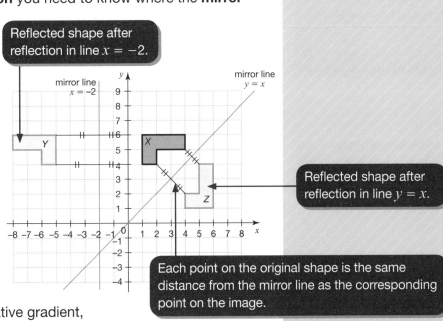

> Reflected shape after reflection in line $y = x$.

> Each point on the original shape is the same distance from the mirror line as the corresponding point on the image.

The diagram shows two reflections of shape X; one reflected in the line $x = -2$ and the other reflected in the line $y = x$.

Rotations

To describe a **rotation**, you need:

- the coordinates of the **centre** of rotation

- the **angle** of rotation (e.g. 90°, 180°, etc.)

- the **direction** of the rotation (clockwise or anticlockwise).

To find these look at the following diagram, where shape A is transformed to shape B. First you have to establish that it is a rotation. Any transformation that is not a reflection or translation must be a rotation.

1. Use a piece of tracing paper and trace shape A.

2. Put a pencil or pen point near to where you think the centre of rotation is.

3. Rotate the tracing paper to see if you can match up the triangle you drew with shape B. If you can't, try another point until the shape matches up.

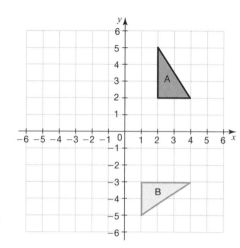

NAILIT!

Never assume that the origin is the centre of rotation as often it is not.

4. Record the angle you moved the tracing paper from its original position, the coordinates of the centre you used and the direction (clockwise or anticlockwise).

Shape A to shape B represents a rotation of 90° clockwise about the point (−1, 0).

Enlargements

To describe an **enlargement**, you need:

- the coordinates of the centre of enlargement

The scale factor can be an integer or a fraction, and positive or negative.

- the scale factor.

The scale factor means how much all the lengths have been multiplied by.

To draw the enlargement of shape A, centre (0, 0) and scale factor 2:

1. Draw lines from the centre of enlargement through the corners of the shape.

2. Choose one vertex of shape A and count how many squares along and up from the centre of enlargement: bottom left vertex is 3 right and 2 up.

A **vertex** is a corner. The plural of vertex is vertices.

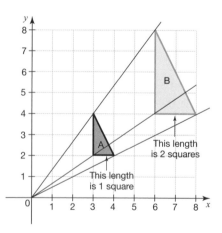

3. Multiply the distance across and up or down by the scale factor. Then plot the corresponding vertex of the image: 3 right × 2 = 6 right; 2 up × 2 = 4 up.

4. Use the scale factor to find the length of each side on the new shape: base of shape A is 1, so the new base will be 1 × 2 = 2.

5. Draw the rest of the shape.

Finding the centre of enlargement and the scale factor

To find the centre and scale factor for an enlargement:

1. Draw a straight line from a vertex on the original shape through the corresponding vertex on the image.

2. Repeat this for two other vertices.

3 The centre of enlargement is where the lines intersect.

4 Use a convenient length on the original shape and measure the corresponding length on the enlarged shape:

scale factor $= \dfrac{\text{length of side on image}}{\text{length of side on object}}$

If the scale factor is greater than 1 then the shape after the enlargement is bigger and if the scale factor is a fraction, then the shape after the enlargement is smaller.

Enlargements with negative scale factors

If the enlargement is the other side of the centre of enlargement, the scale factor is negative.

The enlargement of A to B has scale factor -1 with centre of enlargement (0, 0).

It is called an enlargement even if the shape stays the same size.

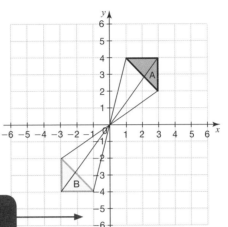

DOIT!

Summarise on a revision card the information you need to give to describe each type of transformation.

CHECKIT!

1 Three coordinates, A(2, 4), B(6, 4) and C(4, 6), are joined to make triangle ABC.

Triangle DEF is an enlargement of triangle ABC with scale factor $\frac{1}{2}$ and centre of enlargement (0, 0).

a Draw triangle ABC on a grid, with x and y from 0 to 8.

b Draw DEF on the same grid.

c Describe the single transformation that maps triangle DEF back on to ABC.

2 Triangle P has vertices (1, 3), (4, 3) and (4, −1).

a Draw triangle P on a grid, with x and y from −8 to 8.

b Reflect triangle P in the line $x = -1.5$.

3 Triangle A has vertices (5, 2), (4, 4) and (7, 2).

a Draw triangle A on a grid, with x and y from −8 to 8.

b Enlarge shape A by scale factor −2 with centre of enlargement (2, 1). Label the image of the enlargement B.

4 Describe the single transformation that maps triangle P onto triangle Q.

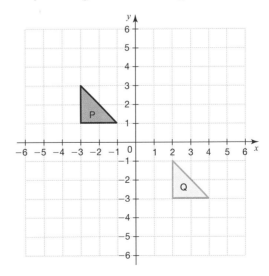

Invariance and combined transformations

Invariance

When combinations of translations, rotations and reflections are applied to a shape, sometimes the positions of all or some of the points on the shape stay the same. These points are said to be **invariant**.

WORKIT!

OABC is a square.

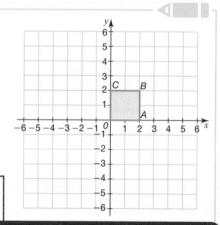

a *OABC* is reflected in the line $x = 2$. How many invariant points on the square are there?

As points A and B lie on the mirror line they will be invariant.

Hence there are two invariant points.

b *OABC* is rotated 90° clockwise about the point A. How many invariant points are there?

There is one invariant point, point A. ◄─────

> All the other points change but point A is invariant as it is the centre of rotation.

c *OABC* is reflected in the line $y = x$. Give the coordinates of the points on the square that are invariant after this reflection.

The invariant points are O (0, 0) and B (2, 2).

> Points O and B lie on the line $y = x$, which is the mirror line, so both these points are invariant.

Combinations of translations, rotations and reflections

When you are told to apply a combination of translations, rotations or reflections to a shape, first transform the original shape and then transform the image to give the final image.

WORKIT!

a Rotate triangle P through 90° clockwise about the origin. Label the image Q.

See diagram on right.

b Reflect image Q in the y-axis. Label the image R.

See diagram on next page.

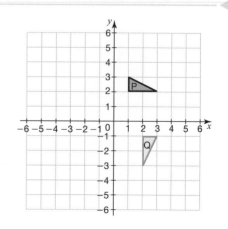

c Describe the single transformation that maps triangle P to triangle R.

Reflection in the line $y = -x$

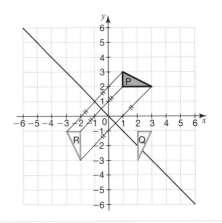

> You can see that R is a reflection of P.
> Notice all the lines from vertices on P to the line $y = -x$ are the same length as the lines from corresponding vertices on R.

DO IT!

Draw a simple shape and perform two translations on it. Describe the single transformation that has the same effect.

✓ CHECK IT!

1 Shape X has vertices (1, 1), (1, 2), (3, 2) and (4, 1).

a Draw shape X on a grid, with x and y from −5 to 5.

b Reflect shape X in the line $y = -1$. Label the image Y.

c Rotate shape Y through 180° about (0, −1). Label the image Z.

d Describe the single transformation that maps shape X to shape Z.

2 Triangle ABC has vertices A(1, 4), B(3, 4) and C(0, 6).

a Draw triangle ABC on a grid, with x and y from 0 to 6.

b Reflect triangle ABC in the line $x = 3$.

c Write down the coordinates of the invariant point during the reflection.

3 The graph below shows shape A.

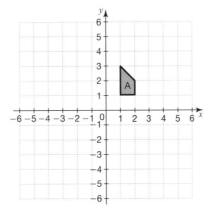

a Rotate shape A 180° about the origin and label it B.

b Shape B is reflected in the y-axis to give shape C. Draw shape C on the graph.

c Describe the single transformation that would be applied to shape C to give shape A.

3D shapes

It is important to know the properties of 3D shapes as you may be asked questions based on these. For example, if you are asked to find the surface area of a square-based pyramid you need to know it has 5 faces: 4 triangles and a square.

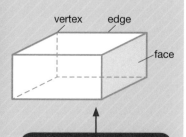

Vertex: a corner (plural is vertices).

Face: a flat exterior surface.

Edge: the line where two faces join.

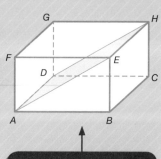

A **plane** is a flat surface. The diagram shows a plane drawn between vertices *A*, *D*, *H* and *E*.

2 flat faces and 1 curved face.

Shape		Faces	Edges	Vertices
Cube		6	12	8
Cuboid		6	12	8
Triangular prism		5	9	6
Hexagonal prism		8	18	12
Cylinder		3	2	0
Square-based pyramid		5	8	5

Shape		Faces	Edges	Vertices
Triangular-based pyramid		4	6	4
Cone		2	1	1
Sphere		1	0	0

A triangular-based pyramid is also called a **tetrahedron**.

The top of a cone is called an apex. This may also be referred to as the vertex.

DO IT!

Work out how many planes of symmetry there are in each of the 3D shapes in this table.

A plane of symmetry is like a line of symmetry for 3D shapes.

CHECK IT!

1 Copy and complete the table.

Shape	Number of vertices	Number of faces	Number of edges
Triangular-based pyramid			
Cone			
Cuboid			
Hexagonal prism			

2 Show that for an octagonal prism, $V + F - E = 2$, where V is the number of vertices, F is the number of faces and E is the number of edges.

Parts of a circle

To answer questions about circles you need to be familiar with all the relevant terms relating to them.

The perimeter of a circle is called the **circumference**. Part of the circumference is called an **arc**.

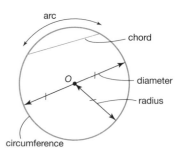

The **radius** of a circle is the distance from the centre (O) to the edge. It is half of the **diameter**.

A **chord** is any straight line across the circle that is not a diameter. A **segment** is any area cut off by a chord.

A **sector** is a pie slice formed by two radii of the circle.

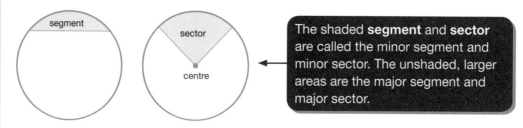

The shaded **segment** and **sector** are called the minor segment and minor sector. The unshaded, larger areas are the major segment and major sector.

DO IT!

Create matching pairs of cards with the terms and diagrams for parts of a circle. Practise matching the cards.

WORKIT!

The perpendicular bisector of a chord of a circle passes through the centre of the circle.

AB is a chord. *OA* and *OB* are the same length as they are both radii of the circle.

Triangle *OAB* is an isosceles triangle so the perpendicular from *O* is a line of symmetry.

The perpendicular bisects the chord: $AM = MB$.

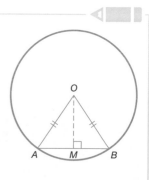

CHECKIT!

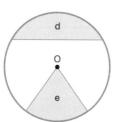

1 Give the names for the labelled parts of a circle (a–e).

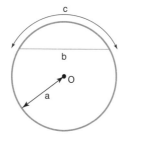

Circle theorems

Circle theorems are used to work out angles in circles. Solving these types of question takes practice in order to identify the theorem you need to use. The key is to be able to choose the relevant theorems when there are lots of angles in the diagram.

SNAP IT!   **Circle theorems**

There are lots of circle theorems to remember. Take a picture of each one to help you revise.

The angle in a semicircle is a right angle.

Make sure that the line you think is the diameter does pass through the centre of the circle.

The angle at the centre is twice the angle at the circumference on the same arc.

Angles at the circumference on the same arc (or equal arcs) are equal.

In the diagram, the lines making both angles start and finish on the same arc.

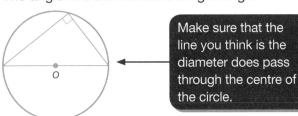

You may need to look carefully to see when to use this. Extra lines drawn on the diagram can make it harder to spot.

Opposite angles of a **cyclic quadrilateral** are **supplementary** (they add up to 180°):
$a + b = 180°$

A cyclic quadrilateral has all four vertices on the circumference of a circle.

Alternate segment theorem: if *AB* is the chord of a circle and a tangent *XY* is drawn at *A,* and *C* is any point on the arc *AB*, then angle *ACB* and angle *BAY* are equal.

The angle between the chord and the tangent equals the angle formed at point C in the other (alternate) segment.

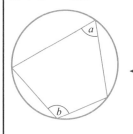

This can be the hardest theorem to spot, so needs practice. In the exam, give the reason as 'alternate segment theorem'.

The tangent at a point on a circle makes a right angle with the radius drawn through the point.

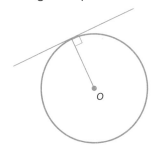

Tangents from the same point to a circle are the same length.

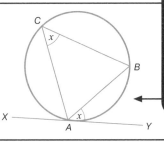

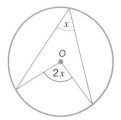

NAIL IT!

As well as using circle theorems, you must apply other geometrical properties. So look out for isosceles triangles, parallel lines, etc.

WORKIT!

A, *B*, *C* and *D* are points on the circumference of a circle with centre *O*. Angle *AOC* = *x*.

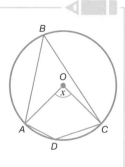

a Find the size of angle *ABC* in terms of *x*.

angle $ABC = \frac{x}{2}$ (angle at the centre is twice the angle at the circumference on the same arc)

b Find the size of angle *ADC* in terms of *x*.

angle ABC + angle $ADC = 180°$
(*ABCD* is a cyclic quadrilateral)

angle $ADC = 180 -$ angle $ABC = \left(180 - \frac{x}{2}\right)°$

WORKIT!

AB is the diameter of the circle. A tangent *XY* to the circle is drawn at point *B*.

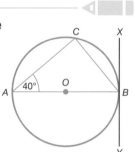

a Write down the size of angle *ACB*, giving a reason for your answer.

angle $ACB = 90°$ (angle in a semicircle is a right angle)

b Write down the size of angle *CBX*, giving a reason for your answer.

angle CBX = angle BAC (alternate segment theorem)
= 40°

c Write down the size of angle *ABY*, giving a reason for your answer.

angle $ABY = 90°$ (angle between tangent and diameter)

CHECKIT!

1 *ABCD* is a cyclic quadrilateral and *O* is the centre of the circle.

Angle *YAB* = 30°

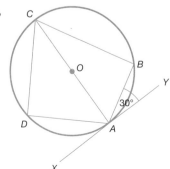

a Work out the size of angle *ACB*, giving a reason for your answer.

b Write down the size of angle *ABC*, giving a reason for your answer.

c Write down the size of angle *ADC*, giving a reason for your answer.

2 A, B and C are points on the circumference of a circle. A tangent AX is drawn at A.

Angle AXO = 30°.

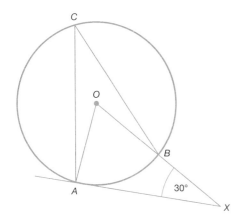

a State the size of angle OAX.

b State, with a reason, the size of angle AOX.

c State, with a reason, the size of angle ACB.

You must give a reason for each stage of your working.

3 A, B, C and D are points on a circle. A tangent is drawn to the circle at D.

Angle ACB = 40° and angle EDA = 50°.

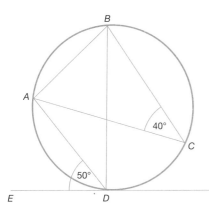

a Write down, with a reason, the size of angle ADB.

b Calculate angle BAD.

You must give a reason for each stage of your working.

4 A, B, C and D are points on the circumference of a circle with centre O.

Angle ADB = 46°

Angle DBC = 50°

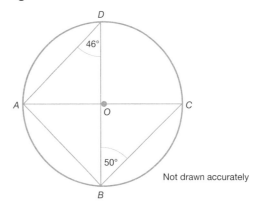

Not drawn accurately

Work out the size of angle BAC giving reasons for your answer.

NAILIT!

Diagrams may be marked 'not to scale'. This means the sizes and proportions shown are not exact – angles or lengths that look similar may not be. Use the information you are given to work them out.

SNAPIT!

Take photos of any questions you see with unusual diagrams, showing your solution, too.

Projections

Projections are views of a solid shape from different directions such as:

- **front elevation** – what you would see directly in front of the solid
- **side elevations** – view from either the left or the right
- **plan** – view from directly above.

You will be told in the question whether both or just one side is required.

In the exam, you will be supplied with the grid.

WORKIT!

The solid shape shown is made up from 11 identical cubes.

Draw on a grid the side elevation from L, the side elevation from R and the plan view.

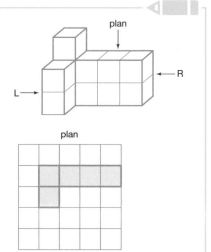

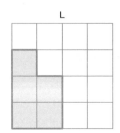
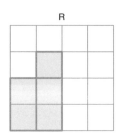

NAILIT!

Work from left to right of the view and shade in the squares as you go.

DOIT!

Build some shapes from cubes and draw two elevations and a plan view of each one.

✓ CHECKIT!

1 This solid shape is made up of 8 identical cubes.

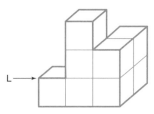

On a grid, draw:

a the side elevation from the left

b the front elevation

c the plan view.

2 The plan, front elevation and side elevation of a prism are drawn on a centimetre grid.

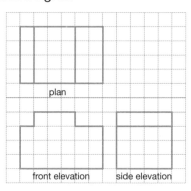

a Draw a sketch of the prism.

b Write the dimensions of the prism on your sketch.

Bearings

Bearings are always:

- measured from due north
- measured in a clockwise direction
- written as three figures (for example, a bearing of 5° would be 005° and a bearing of 32.5° would be 033°).

East is a bearing of 090°, south 180° and west 270°.

NAILIT!

You must be able to draw lines in the compass directions north, east, south and west.

NAILIT!

You often need to look for alternate angles when you draw diagrams showing bearings between two or more points.

NAILIT!

The 'from' part in the question is important because it tells you where you are starting from. 'From point A' means you should measure the bearing from this point.

WORKIT!

> Draw a diagram.

Point B is on a bearing of 100° from point A.

What is the bearing of point A from point B?

From the diagram, $\theta = 180 + 100 = 280°$

The bearing of point A from point B is 280°

> The bearing from B to A is the clockwise angle θ.

> North lines are always parallel so you can use these to identify alternate angles.

NAILIT!

Add north lines through all the points and mark any angles mentioned in the question that are not included on the diagram.

NAILIT!

If the question does not include a diagram, draw one to help you understand the problem.

DOIT!

Describe a bearing in no more than 12 words.

✓ CHECKIT!

1 Ship S sets off on a straight course of 245° for port P.

What is the bearing of P from S?

2 The diagram shows the positions of three wind turbines, P, Q and R.

The bearing of Q from P is 045°.

The bearing of R from Q is 140°.

a Calculate the bearing of P from Q.

b Calculate the bearing of Q from R.

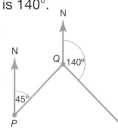

Pythagoras' theorem

Memorise: the square on the hypotenuse is equal to the sum of the squares on the other two sides.

Pythagoras' theorem states that the square of the hypotenuse of any right-angled triangle is equal to the lengths of the other two sides squared and added together.

SNAP IT! Pythagoras' theorem

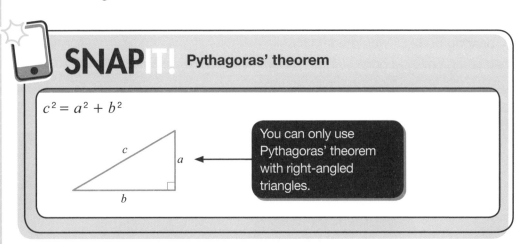

$c^2 = a^2 + b^2$

You can only use Pythagoras' theorem with right-angled triangles.

NAIL IT!

You need to learn the formula for Pythagoras' theorem, as you will not be given it in the exam.

WORKIT!

An isosceles triangle has base 8 cm and the pair of equal sides of length 10 cm. Find, to 3 significant figures

a the perpendicular height of the triangle

Using Pythagoras' theorem $10^2 = h^2 + 4^2$

$$100 = h^2 + 16$$

$$84 = h^2$$

$$h = \sqrt{84} = 9.17 \text{ cm (to 3 s.f.)}$$

10 cm h

4 cm

b the area of the triangle.

$$\text{Area} = \frac{1}{2} \times \text{base} \times \text{height}$$

$$= \frac{1}{2} \times 8 \times 9.17$$

$$= 36.7 \text{ cm}^2 \text{ (to 3 s.f.)}$$

Draw a diagram.
The perpendicular height of an isosceles triangle cuts the base into two equal lengths.

NAIL IT!

The area of a triangle is covered on page 152.

NAIL IT!

Pythagoras' theorem can occur in many questions on different topics, for example finding the distance between two points whose coordinates are given.

DO IT!

Work out a formula for finding the distance between two points whose coordinates are given.

SNAP IT!

Take photos of right-angled triangles you see around you, and use them to create your own questions.

✓ CHECK IT!

1 Find length x in these triangles, giving your answers to 1 decimal place.

a

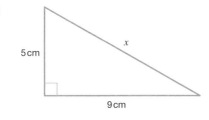

5 cm

9 cm

x

b

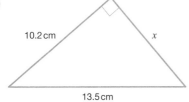

10.2 cm

x

13.5 cm

2 The lengths of the sides of an isosceles triangle are 12 cm, 12 cm and 10 cm. Find the perpendicular height of the triangle and hence its area, giving your answer to 2 decimal places.

3 Triangles ABC and ACD are right-angled triangles.

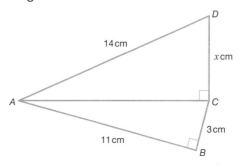

D

14 cm

x cm

A

C

3 cm

11 cm

B

Work out the value of x, correct to 2 decimal places.

4 The isosceles triangle shown below has base $AB = 14$ cm.

The area of the triangle ABC is 90 cm².

Calculate the length of AC giving your answer to 3 significant figures.

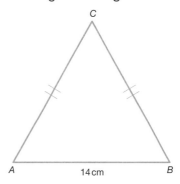

C

A   14 cm   B

Area of 2D shapes

SNAP IT!   Area and volume

There are lots of formulae to remember for area and volume. Take a picture of each one to help you revise.

Area of triangle	Area of parallelogram	Area of trapezium	Circumference of circle
$= \frac{1}{2} \times$ base $\times$ perpendicular height	$=$ base $\times$ perpendicular height	$= \frac{1}{2}(a + b)h$	$= 2\pi r = \pi D$ Area of circle $= \pi r^2$

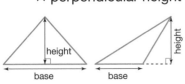
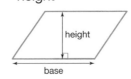
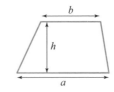

DO IT!

Write each formula and its shape on pairs of cards. Practise matching the cards.

NAIL IT!

The Greek letter π equals the area of a circle with radius 1 unit and has the value 3.14159…. It is an irrational number (a non-repeating, non-terminating decimal). Check that you know which key to use on your calculator.

CHECK IT!

1 A wooden floor is going to be put in the office shown by this plan. The floor costs £38 per m². How much will the new floor cost?

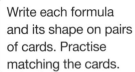

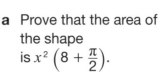

2 The diagram shows a square with a perimeter of 20 cm.

Without using a calculator, work out the shaded area as a proportion of the area of the square. Give your answer as a fraction.

3 Shape *ABCDE* consists of a semicircle with diameter *BD* and a rectangle *ABDE*.

$BD = 2x$ and $AB = 4x$.

a Prove that the area of the shape is $x^2 \left(8 + \frac{\pi}{2} \right)$.

b Find the exact perimeter of the shape *ABCDE* in terms of x.

Volume and surface area of 3D shapes

SNAP IT! Volume

There are lots of formulae to remember for area and volume. Take a picture of each one to help you revise.

NAIL IT!

Be careful with boxes as sometimes they don't have a top. You will need to take this into account when finding the surface area.

Volume

Volume of cuboid = length × width × height	Volume of prism = area of cross-section × length
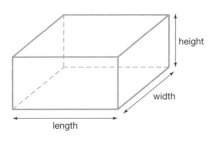	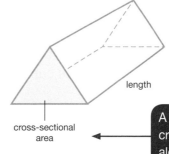 A prism has the same cross-sectional shape along its length.
Volume of cylinder = $\pi r^2 h$ 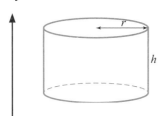 A cylinder is a prism with a circular cross-section so the area can be found by multiplying the cross-sectional area by the height/length.	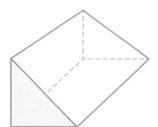
Volume of a cone = $\frac{1}{3}\pi r^2 h$ 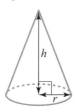	Volume of sphere = $\frac{4}{3}\pi r^3$ 

WORKIT!

1 *ABCDEF* is a prism with an equilateral triangle cross-section of side 3 cm. The length of the prism is 12 cm.

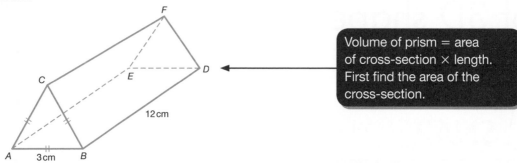

Volume of prism = area of cross-section × length. First find the area of the cross-section.

Work out the volume of the prism. Give your answer to 2 significant figures.

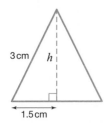

Using Pythagoras' theorem, $3^2 = h^2 + 1.5^2$

$9 = h^2 + 2.25$

$h = 2.60$ cm (to 2 d.p.)

To find the area of the triangle you need to find the perpendicular height, h, first.

Area of triangle $= \frac{1}{2} \times$ base $\times$ height

$= \frac{1}{2} \times 3 \times 2.60 = 3.9$ cm² (to 2 s.f.)

Volume of prism = area of cross-section × length

$= 3.9 \times 12 = 47$ cm³ (to 2 s.f.)

2 100 steel ball bearings of radius 1 cm are melted down and made into a cuboid with a base area of 60 cm². Calculate the height of the cuboid in cm, giving your answer correct to 3 significant figures.

Volume of one ball bearing $= \frac{4}{3}\pi r^3 = \frac{4}{3}\pi \times 1^3 = 4.1888$ cm³

First find the volume of one sphere.

Volume of 100 ball bearings $= 100 \times 4.1888 = 418.88$ cm³

Volume of cuboid = area of base × height

A cuboid is an example of a prism with a rectangular cross-section.

$418.88 = 60 \times h$

$h = \frac{418.88}{60}$ cm $= 6.98$ cm (to 3 s.f.)

Surface area

Curved surface area of cylinder	Curved surface area of cone = $\pi r l$	Surface area of sphere
$= 2\pi r h$		$= 4\pi r^2$
Total surface area of cylinder	Total surface area of cone = $\pi r l + \pi r^2$	
$= 2\pi r h + 2\pi r^2$		

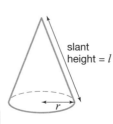

Note that the formula for the curved surface area uses the slant height, but the volume uses the perpendicular height.

slant height = l

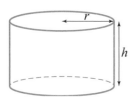

STRETCHIT!

Sometimes you may be given the perpendicular height and the radius. You need to use Pythagoras' theorem to find the slant height: $r^2 + h^2 = l^2$.

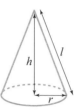

DOIT!

Write each formula and its shape on pairs of cards. Practise matching the cards.

WORKIT!

A solid sphere of radius 30 cm is cut in half to form two identical solids. Find the surface area of one of the solids, giving your answer in terms of π.

Curved surface area of hemisphere $= \dfrac{4\pi r^2}{2} = 2\pi r^2 = 2 \times \pi \times 30^2$

$\qquad\qquad\qquad\qquad\qquad\qquad = 1800\pi \text{ cm}^2$

Area of circular base $\pi r^2 = \pi \times 30^2 = 900\pi \text{ cm}^2$

Total surface area of solid $= 1800\pi + 900\pi = 2700\pi \text{ cm}^2$

The surface area is the curved surface area of the hemisphere plus the area of the circular base.

Remember to leave answers in terms of π.

CHECKIT!

1 A cylinder of radius a m and height $6a$ m is reformed into a thin cylindrical rod with radius r m and length 4 m. Write an expression for the radius of the rod, r, in terms of a.

2 The diagram shows the cross-section of a rectangular swimming pool.

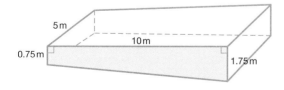

a Find the cross-sectional area shaded in the diagram.

b Given that the swimming pool is a prism, calculate the volume of water in the pool in m³.

The swimming pool is filled using two hoses. Each hose supplies a volume of water at 0.05 m³ per minute.

c Calculate the number of hours it would take to fill the pool with water. Give your answer to the nearest hour.

3 A circus tent is in the shape of a cone on a cylinder. The radius of the base is 15 m, the height of the cylinder is 5 m and the perpendicular height of the cone is 4 m.

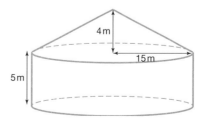

Find the surface area of fabric used for the tent, correct to 3 significant figures. Assume that the tent has no floor.

Trigonometric ratios

SNAPIT! Trigonometric ratios

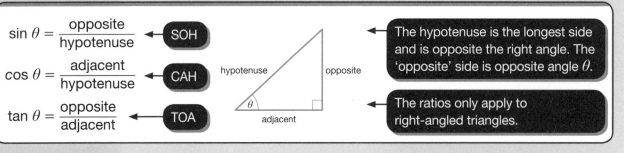

$$\sin \theta = \frac{\text{opposite}}{\text{hypotenuse}} \leftarrow \text{SOH}$$

$$\cos \theta = \frac{\text{adjacent}}{\text{hypotenuse}} \leftarrow \text{CAH}$$

$$\tan \theta = \frac{\text{opposite}}{\text{adjacent}} \leftarrow \text{TOA}$$

The hypotenuse is the longest side and is opposite the right angle. The 'opposite' side is opposite angle θ.

The ratios only apply to right-angled triangles.

NAILIT!

These ratios will not be given in the exam so you need to remember them.

Use the mnemonic SOH CAH TOA to help.

Finding a length

Here's how to use the trigonometric ratios to find an unknown length in a right-angled triangle when one side and one angle are known.

1. Identify which side is known.

2. Work out which trigonometric ratio to use from SOH CAH TOA.

3. Write the trigonometric ratio and substitute the values.

4. Solve the equation for the unknown length.

WORKIT!

Find the length of the side marked x, giving your answer to 3 significant figures.

$$\tan \theta = \frac{opposite}{adjacent}$$

The sides are O and A so use TOA.

Write the values in $\tan \theta = \frac{\text{opposite}}{\text{adjacent}}$ x is in the denominator of the fraction.

$$\tan 30° = \frac{6}{x}$$

$$x \tan 30° = 6$$

$$x = \frac{6}{\tan 30°} = 10.4 \, cm \text{ (to 3 s.f.)}$$

Finding an angle

Similarly, you can use the trigonometric ratios to find an angle of a right-angled triangle when the lengths of two sides are known. First find the sin, cos or tan. Then find the angle.

WORKIT!

1 Find the size of the angle θ, giving your answer to 1 decimal place.

$\cos\theta = \dfrac{\text{adjacent}}{\text{hypotenuse}}$ ← The sides are A and H so use CAH.

$= \dfrac{8}{10} = 0.8$

$\theta = \cos^{-1}(0.8)$

$= 36.9°$ (to 1 d.p.)

2 Triangle *ABC* has an area of 25 cm². Side *AB* is 10 cm.

a Find the perpendicular height, h.

Area $= \dfrac{1}{2} \times$ base $\times$ perpendicular height

$25 = \dfrac{1}{2} \times 10 \times h$

$h = 5\text{cm}$

b Calculate the length *AC*.

$\sin 30° = \dfrac{5}{AC}$ ← Substitute values into $\sin\theta = \dfrac{\text{opposite}}{\text{hypotenuse}}$.

$AC = \dfrac{5}{\sin 30°} = \dfrac{5}{0.5} = 10\text{ cm}$

NAILIT!

Check that you know how to find inverse sine, cosine and tangent on your calculator. The key may be shown as

$\boxed{\sin^{-1}}$ or

$\boxed{\text{arcsin}}$

Solving problems in 3D

To solve 3D problems, you need to identify right-angled triangles and then use these with Pythagoras' theorem and the trigonometric ratios.

WORKIT!

The diagram shows the internal dimensions of a rectangular box.

The longest diagonal will be one of the lengths like *AD*.

a Find the length of the longest straight rod that will just fit inside the box.

Use Pythagoras' theorem with triangle *ABC* to find *AC²*.

$AC^2 = 60^2 + 30^2 = 4500$

$AD^2 = 4500 + 20^2 = 4900$

$AD = \sqrt{4900} = 70\text{ cm}$

Longest length of rod is 70 cm.

Angle *ACD* is 90° so use Pythagoras theorem with triangle *ACD* to find *AD*.

b Work out the angle that the rod makes with the base of the box.

Angle *CAD* is the angle with the base of the box.

Let angle $CAD = \theta$

$\sin\theta = \dfrac{\text{opposite}}{\text{hypotenuse}} = \dfrac{20}{70} = 0.2857$

Use sides *CD* and *AD* with sin.

$\theta = \sin^{-1}(0.2857) = 16.6°$ (to 1 d.p.)

The rod makes an angle of 16.6° with the base of the box.

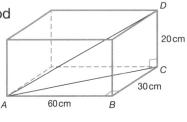

NAILIT!

In 3D questions, you may find it helpful to draw just the triangle you are working on.

DOIT!

Look up different mnemonics for remembering the ratios. Choose one you find easy to remember.

CHECKIT!

1 Find the lengths of the sides marked x. Give your answers to 2 decimal places.

a

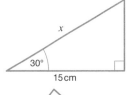

x

$30°$

$15\,\text{cm}$

b

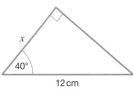

x

$40°$

$12\,\text{cm}$

2 Find angle θ for each of these triangles. Give your answers to the nearest $0.1°$.

a

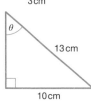

$10\,\text{cm}$

θ

$3\,\text{cm}$

b

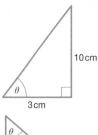

θ

$13\,\text{cm}$

$10\,\text{cm}$

3 In terms of a, b and c, prove that for $0° < \theta < 90°$, $\dfrac{\sin\theta}{\cos\theta} = \tan\theta$.

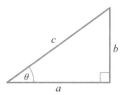

c

b

θ

a

4 The diagram shows a cuboid box with edges $AB = 10\,\text{cm}$, $AD = 7\,\text{cm}$ and $BC = 6\,\text{cm}$.

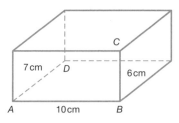

C

$7\,\text{cm}$ D $6\,\text{cm}$

A $10\,\text{cm}$ B

a Calculate the length of CD, correct to 2 decimal places.

b Work out angle BCD, correct to 1 decimal place.

c Point P is the midpoint of BD. Calculate the length of PC, correct to 2 decimal places.

Exact values of sin, cos and tan

You already know the values of the sine and cosine of 0° and 90° and of tan 0° from the graphs.

You can find the graphs of sine, cosine and tangent on pages 84–5.

The values for 30°, 45° and 60° can be determined by drawing triangles, working out the lengths of the sides that aren't known and then using trigonometry.

Sine, cosine and tangent of 45°

The exact values can be worked out by drawing a right-angled triangle with angles of 45° and short sides of 1 unit.

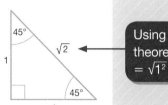

Using Pythagoras' theorem, hypotenuse $= \sqrt{1^2 + \sqrt{1^2}} = \sqrt{2}$

$$\sin 45° = \frac{\text{opposite}}{\text{hypotenuse}} = \frac{1}{\sqrt{2}}$$

Substitute the lengths of the sides in the trigonometric ratios.

$$\cos 45° = \frac{\text{adjacent}}{\text{hypotenuse}} = \frac{1}{\sqrt{2}}$$

$$\tan 45° = \frac{\text{opposite}}{\text{adjacent}} = \frac{1}{1} = 1$$

NAILIT!

If you are asked to find the exact value use surds instead of decimals.

These are all exact values. For example, entering $\cos 45°$ into a calculator gives 0.70710657812…, which cannot be expressed as a fraction. In calculations where answers are given to a number of decimal places or significant figures, the answers are not exact values. Numbers expressed in terms of surds are exact.

Sine, cosine and tangent of 30° and 60°

These exact values can be worked out by using an equilateral triangle of side length 2. Bisecting one of the sides then gives a right-angled triangle (A).

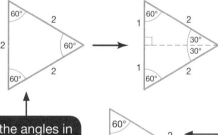
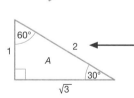

Bisecting the triangle gives a right-angled triangle with angles 60° and 30°.

Using Pythagoras' theorem, base $= \sqrt{2^2 - 1^2} = \sqrt{3}$

$$\sin 30° = \frac{1}{2} \qquad \sin 60° = \frac{\sqrt{3}}{2}$$

$$\cos 30° = \frac{\sqrt{3}}{2} \qquad \cos 60° = \frac{1}{2}$$

$$\tan 30° = \frac{1}{\sqrt{3}} \qquad \tan 60° = \sqrt{3}$$

All the angles in the equilateral triangle are 60°.

DOIT!

Draw a table of the values of sin, cos and tan of 0°, 30°, 45°, 60° and 90°.

CHECKIT!

1 Show that $\sqrt{3} \tan 30° + \cos 60°$ can be written as $\frac{a}{b}$, where a and b are integers.

2 Show that $\sin^2 30° + \cos^2 30° = 1$.

Sectors of circles

SNAPIT! Sectors

For a sector of radius r and angle θ:

- length of arc: $l = \dfrac{\theta}{360} \times 2\pi r$

- area of sector: $A = \dfrac{\theta}{360} \times \pi r^2$

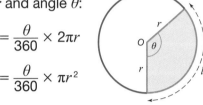

You can work out the arc length of a sector of angle θ degrees by multiplying the circumference of the circle by $\dfrac{\theta}{360}$. Similarly, you can work out the area of a sector of angle θ degrees by multiplying the area of the circle by $\dfrac{\theta}{360}$.

WORKIT!

Judy is planning a wedding to be attended by 57 guests. The tables have a diameter of 2.3 m. Each guest needs a distance of 70 cm along an arc around the table.

Calculate the maximum number of guests that can be seated at each table.

Circumference of table $= \pi D = \pi \times 2.3 = 7.23\,m$

$7.23\,m = 723\,cm$

> Make sure that both numbers are in the same units.

Number of guests at each table $= \dfrac{723}{70} = 10.3$

Maximum number at a table $= 10$

> The number of guests must be a whole number.

WORKIT!

The diagram shows a sector with radius 6 cm. Its area is 24 cm².

a Work out angle x, giving your answer to 1 decimal place.

> Write down the formula for the area of a sector.

$A = \dfrac{x}{360}\pi r^2$

$24 = \dfrac{x}{360} \times \pi \times 6^2$

> Substitute the values you know.

> Cancel this before calculating.

$x = \dfrac{24 \times 360}{\pi \times 36} = 76.4°$ (to 1 d.p.)

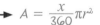

b Calculate the length of the arc, l, giving your answer to the nearest whole number.

> Write down the formula for the area of a sector.

$l = \dfrac{x}{360} \times 2\pi r = \dfrac{76.4}{360} \times 2\pi \times 6 = 8\,cm$ (to nearest whole number)

DO**IT!**

Make a revision poster for the length of an arc and the perimeter and area of a sector.

CHECK**IT!**

1 *AB* is the arc of a circle of radius 12 cm. The length of the arc *AB* is 10 cm.

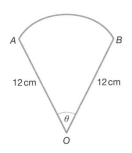

Calculate the size of angle θ to 1 decimal place.

2 This logo is made of two identical sectors and an isosceles triangle. The sectors have radius 2 cm and angle 40°. The base of the isosceles triangle is 1.2 cm.

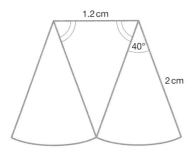

Work out

a the perimeter of the logo

b the area of the logo.

Give your answers to 3 significant figures.

3 The area of sector *OAB* is 25π cm².

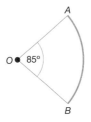

Find the radius *OA* to 1 decimal place.

4 A circular piece of paper has a sector removed to leave the shaded shape shown below.

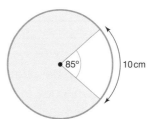

Calculate the area of the shaded shape. Give your answer to the nearest whole number.

Sine and cosine rules

The **sine rule** and **cosine rule** can be used with any triangle, not just those containing a right angle.

The angles are denoted by the letters A, B and C and the lengths of the sides opposite these angles by a, b and c.

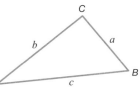

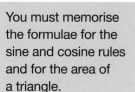

NAILIT!

You must memorise the formulae for the sine and cosine rules and for the area of a triangle.

SNAPIT! **Sine and cosine rules**

The sine rule: $\dfrac{a}{\sin A} = \dfrac{b}{\sin B} = \dfrac{c}{\sin C}$

The cosine rule: $a^2 = b^2 + c^2 - 2bc \cos A$

DOIT!

Using the diagram above and what you already know about triangles, see whether you can prove that these rules are true.

The area of a triangle

If you don't know the height of a triangle but you do know the lengths of two sides and the included angle, you can find its area using the formula below.

SNAPIT! **Area of any triangle**

Area of triangle $= \dfrac{1}{2} ab \sin C$

This formula works for all triangles, but for areas of right-angled triangles, use the formula $A = \frac{1}{2} \times$ base $\times$ height.

Warning: be careful when using this formula to find the angle from the area and two sides of a triangle. For example, $\sin C = \frac{1}{2}$ can have two solutions: 30° and 150° so the triangle could be obtuse-angled. If the triangle has been drawn or other angles in the triangle are known, then the obtuse angle might not be a possible solution. Check the question carefully.

WORKIT!

1 In triangle PQR, $PR = 8\,$cm, $QR = 12\,$cm and angle $PRQ = 150°$.

   a Find the area of triangle PQR.

Always write out the formula and then substitute the numbers you know.

$\text{Area} = \frac{1}{2} ab \sin C = \frac{1}{2} \times 12 \times 8 \times \sin 150° = 24\,\text{cm}^2$

b Find the length of *PQ*, correct to 1 decimal place.

Using the cosine rule:

$a^2 = b^2 + c^2 - 2bc\cos A$

$= 12^2 + 8^2 - 2 \times 12 \times 8 \times \dfrac{-\sqrt{3}}{2}$ ← $\text{Cos}150° = -\cos30°$

$= 374.2769$

$a = 19.3462$

$PQ = 19.3\,\text{cm}$ (to 1 d.p.)

NAILIT!

Don't round off answers to the required number of decimal places until the final answer.

2 In triangle *ABC*, *AB* = 8 cm, *BC* = 15 cm and angle *ABC* = 60°.

a Calculate the length of side *AC*.

Using the cosine rule:

$a^2 = b^2 + c^2 - 2bc\cos A$

$AC^2 = 8^2 + 15^2 - 2 \times 8 \times 15 \times \cos60°$

$= 64 + 225 - 240 \times \dfrac{1}{2}$

$= 169$

$AC = 13\,\text{cm}$

b Find the size of angle θ, giving your answer to 1 decimal place.

Using the sine rule:

$\dfrac{b}{\sin B} = \dfrac{c}{\sin C}$

$\dfrac{13}{\sin 60°} = \dfrac{8}{\sin\theta}$

$\sin\theta = \dfrac{8 \times \sin60°}{13}$

$\theta = 32.2°$ (to 1 d.p.)

DOIT!

Draw a right-angled triangle with sides a, b and c. What happens when you apply the sine and cosine rules and the formula for the area of a triangle to this triangle?

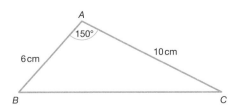

CHECKIT!

1 In triangle *ABC*, *AB* = 6 cm, *AC* = 10 cm and *BAC* = 150°.

a Find the area of triangle *ABC*.

b Find the length of *BC*, correct to 3 significant figures.

2 A triangle has sides 5 cm, 7 cm and 10 cm. Calculate the smallest angle of the triangle, to the nearest 0.1°.

3 In triangle *XYZ*, angle *XZY* is obtuse.

XY = 9.5 cm and *YZ* = 5.4 cm.

The area of the triangle is 16 cm².

a Find angle *XYZ*, correct to 1 decimal place.

b If you did not know that angle *XZY* was obtuse, how would this affect your answer to part a?

Vectors

Vectors are quantities that have both a **size** and a **direction**. There are a few ways to write them:

- small letters or combinations of small letters written in bold (e.g. 2**a**, **a** + **b**, 3(**a** − **b**))

- two letters with an arrow over them (e.g. $\overrightarrow{OP}$) showing the starting and finishing points of the line, together with the direction (O to P in this case)

- a column vector (e.g. $\begin{pmatrix} 3 \\ -1 \end{pmatrix}$ which means 3 units horizontally to the right and 1 unit down or $\begin{pmatrix} -4 \\ 5 \end{pmatrix}$ which means 4 units horizontally to the left and 5 units up).

If $\overrightarrow{OA}$ = **a** and $\overrightarrow{OB}$ = **b**, then the vector from A to B is found by going from A to O and then from O to B. The vector A to O is minus the vector from O to A.

This can be written as $\overrightarrow{AB} = \overrightarrow{AO} + \overrightarrow{OB}$ or −**a** + **b** = **b** − **a**.

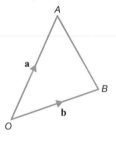

If one vector is **a** and the other is −**a**, then the two vectors are the same length but opposite in direction.

If two vectors are parallel, they have the same vector part, for example **a** and 2**a**. The numbers in front are called **scalars** and show how long the vector is. 2**a** is therefore twice as long as **a**, and $\frac{2}{3}$**b** is $\frac{2}{3}$ the length of **b**.

2**a** + 2**b** can be written as 2(**a** + **b**). This vector is parallel to and twice the length of **a** + **b**.

The table gives further examples.

1st vector	2nd vector	Direction?	Length?	Parallel?
4**a**	4**a**	Same	Same	Yes
2**a** + **b**	4**a** + 2**b**	Same	2nd vector is twice as long	Yes
4**a** − **b**	4**a** + **b**	Not the same	Not the same	No
3(**b** − **a**)	2(**b** − **a**)	Same	1st vector is longer with lengths in ratio of 3:2	Yes
−4**a** + 2**b**	**b** − 2**a**	Same	1st vector is twice as long	Yes
−(3**a** − **b**)	3**a** − **b**	Opposite	Same	Yes
$\frac{1}{3}$(**a** − **b**)	**b** − **a**	Opposite	2nd vector is three times as long	Yes

Calculating with column vectors

You can calculate with column vectors:

- To add $\begin{pmatrix} 3 \\ 2 \end{pmatrix}$ and $\begin{pmatrix} 4 \\ -1 \end{pmatrix}$, add the separate elements:

$$\begin{pmatrix} 3 \\ 2 \end{pmatrix} + \begin{pmatrix} 4 \\ -1 \end{pmatrix} = \begin{pmatrix} 3+4 \\ 2+(-1) \end{pmatrix} = \begin{pmatrix} 7 \\ 1 \end{pmatrix}$$

- To subtract $\begin{pmatrix} 4 \\ -1 \end{pmatrix}$ from $\begin{pmatrix} 3 \\ 2 \end{pmatrix}$, subtract the separate elements:

$$\begin{pmatrix} 3 \\ 2 \end{pmatrix} - \begin{pmatrix} 4 \\ -1 \end{pmatrix} = \begin{pmatrix} 3-4 \\ 2-(-1) \end{pmatrix} = \begin{pmatrix} -1 \\ 3 \end{pmatrix}$$

- To multiply a vector by a number, multiply each element by the number:

$$5\begin{pmatrix} -2 \\ 4 \end{pmatrix} = \begin{pmatrix} 5 \times (-2) \\ 5 \times 4 \end{pmatrix} = \begin{pmatrix} -10 \\ 20 \end{pmatrix}$$

WORKIT!

$a = \begin{pmatrix} -5 \\ 3 \end{pmatrix}$ and $b = \begin{pmatrix} 7 \\ 1 \end{pmatrix}$

Work out as a column vector

a $a + 2b$

$$a + 2b = \begin{pmatrix} -5 \\ 3 \end{pmatrix} + 2\begin{pmatrix} 7 \\ 1 \end{pmatrix} = \begin{pmatrix} -5 + 2 \times 7 \\ 3 + 2 \times 1 \end{pmatrix} = \begin{pmatrix} 9 \\ 5 \end{pmatrix}$$

b $2a - 3b$

$$2a - 3b = 2\begin{pmatrix} -5 \\ 3 \end{pmatrix} - 3\begin{pmatrix} 7 \\ 1 \end{pmatrix} = \begin{pmatrix} 2 \times (-5) - 3 \times 7 \\ 2 \times 3 - 3 \times 1 \end{pmatrix} = \begin{pmatrix} -31 \\ 3 \end{pmatrix}$$

Working with vectors

WORKIT!

> Make sure that you use vector $\overrightarrow{PO}$ not $\overrightarrow{OP}$.

1 OPQ is a triangle. M is the midpoint of PQ.
$\overrightarrow{OP} = a$ and $\overrightarrow{OQ} = b$.

a Express $\overrightarrow{PQ}$ in terms of a and b.

$\overrightarrow{PQ} = \overrightarrow{PO} + \overrightarrow{OQ}$

$= -a + b$ ◄ $\overrightarrow{PO} = -\overrightarrow{OP}$.

$= b - a$

b Express $\overrightarrow{OM}$ in terms of a and b, giving your answer in its simplest form.

$\overrightarrow{PM} = \frac{1}{2}\overrightarrow{PQ} = \frac{1}{2}(b - a)$ ◄ M is the midpoint of PQ so $\overrightarrow{PM} = \frac{1}{2}\overrightarrow{PQ}$.

$\overrightarrow{OM} = \overrightarrow{OP} + \overrightarrow{PM}$

$= a + \frac{1}{2}(b - a)$

$= a + \frac{1}{2}b - \frac{1}{2}a$

$= \frac{1}{2}(b + a)$

DOIT!

Write some column vectors on revision cards. Pick two or three cards at a time and make a question from them.

NAILIT!

If it is possible, always take a factor outside the bracket. This helps you to spot any parallel vectors/lines.

2 *OAB* is a triangle with $\overrightarrow{OA} = \boldsymbol{a}$ and $\overrightarrow{OB} = \boldsymbol{b}$.

a M is the midpoint of $\overrightarrow{AB}$ and *Q* is the midpoint of $\overrightarrow{OA}$. Find $\overrightarrow{QM}$.

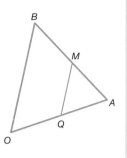

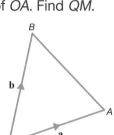

M is the midpoint of $\overrightarrow{AB}$, so find vector $\overrightarrow{AB}$ in terms of **a** and **b**.

$\overrightarrow{AB} = \overrightarrow{AO} + \overrightarrow{OB} = -\boldsymbol{a} + \boldsymbol{b} = \boldsymbol{b} - \boldsymbol{a}$

$\overrightarrow{AM} = \frac{1}{2}\overrightarrow{AB} = \frac{1}{2}(\boldsymbol{b} - \boldsymbol{a})$ ◄— *M* is the midpoint of $\overrightarrow{AB}$ so $\overrightarrow{AM} = \frac{1}{2}\overrightarrow{AB}$.

$\overrightarrow{QA} = \frac{1}{2}\overrightarrow{OA} = \frac{1}{2}\boldsymbol{a}$

$\overrightarrow{QM} = \frac{1}{2}\overrightarrow{OA} + \overrightarrow{AM} = \frac{1}{2}\boldsymbol{a} + \frac{1}{2}(\boldsymbol{b} - \boldsymbol{a})$

$\qquad = \frac{1}{2}\boldsymbol{a} + \frac{1}{2}\boldsymbol{b} - \frac{1}{2}\boldsymbol{a}$

$\qquad = \frac{1}{2}\boldsymbol{b}$

b Using your answer to part a, prove that *QM* and *OB* are parallel.

$\overrightarrow{QM}$ and $\overrightarrow{OB}$ have the same vector part, **b**, so they must be parallel.

The same vector part means the vectors are parallel and the scalar shows that $\overrightarrow{QM}$ is half the length of $\overrightarrow{OB}$.

CHECK**IT!**

1 $\boldsymbol{a} = \begin{pmatrix} 4 \\ -5 \end{pmatrix}$ and $\boldsymbol{b} = \begin{pmatrix} -2 \\ 3 \end{pmatrix}$

Work out as a column vector

a $\boldsymbol{b} - \boldsymbol{a}$ **b** $3\boldsymbol{a} + 5\boldsymbol{b}$

2 *OAB* is a triangle with $\overrightarrow{OA} = \boldsymbol{a}$ and $\overrightarrow{OB} = \boldsymbol{b}$.

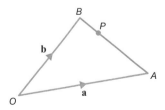

a Find $\overrightarrow{AB}$ in terms of **a** and **b**.

P is the point on *AB* such that $AP:PB = 3:2$.

b Find $\overrightarrow{AP}$ in terms of **a** and **b**.

c Find $\overrightarrow{OP}$ in terms of **a** and **b**.

3 *ABC* is a triangle and *ABD* is a straight line.

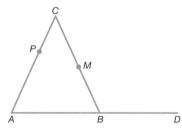

$\overrightarrow{AB} = \boldsymbol{a}$, $\overrightarrow{CP} = \boldsymbol{b}$ and $\overrightarrow{PA} = 2\boldsymbol{b}$.

a Find the vector $\overrightarrow{BC}$ in terms of **a** and **b**.

b *M* is the midpoint of *BC* and *B* is the midpoint of *AD*.

Prove that *PMD* is a straight line.

1 A, B, C and D are points on the circumference of a circle. EDF is a tangent to the circle at D.

Angle $ABC = x°$ and angle $ACD = y°$.

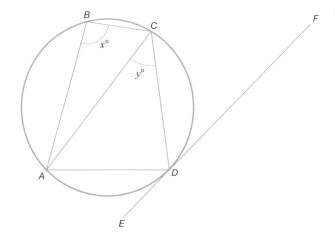

Work out angle CDF in terms of x and y. Give reasons for every stage of your working.

2 Show that $\cos 45° + \sin 60° = \frac{1}{2}(\sqrt{2} + \sqrt{3})$.

3 A right-angled triangle has sides $4x - 3$, $x + 1$ and $3x$.

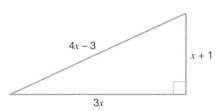

Use Pythagoras' theorem to find the value of x and hence find the lengths of the sides of the triangle.

4 The diagram shows a solid shape made up of seven identical one-centimetre cubes.

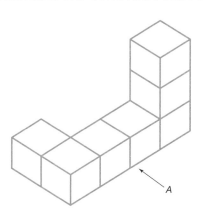

On centimetre grids, draw

a the elevation from A

b the plan view.

5 Triangle X has vertices (2, 2), (4, 3) and (5, 2).

a Draw triangle X on a grid, with x and y from -6 to 6.

b **i** Reflect triangle X in the line $y = x$. Label the image Y.

  ii Write down the coordinates of the invariant point during the reflection.

c Reflect triangle Y in the line $y = 1$. Label the image Z.

d Describe the single transformation that maps triangle X to triangle Z.

6 A ship travels 15 km from A on a bearing of 070°. At B, it travels 10 km on a bearing of 150° to reach C.

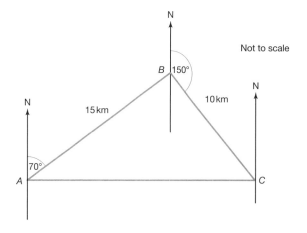

Work out

a the distance from A to C, to 3 s.f.

b the bearing of A from C to the nearest degree.

Probability

The basics of probability

The probability scale

The probability scale goes from 0 to 1. Probabilities can be expressed as decimals or fractions.

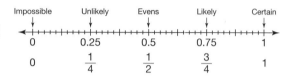

The probability formula

If all the outcomes are equally likely to happen:

$$\text{Probability} = \frac{\text{number of ways something can happen}}{\text{total number of possible outcomes}}$$

NAILIT!

Unbiased or **fair** means that all the outcomes are equally likely. The opposite is **biased**, where certain outcomes would be more likely than others.

WORKIT!

Find the probability of throwing an even number on an unbiased dice.

$$\text{Probability} = \frac{\text{number of ways something can happen}}{\text{total number of possible outcomes}} = \frac{3}{6} = \frac{1}{2}$$

Number of even numbers = 3 (2, 4 and 6).

Total number of numbers = 6.

Always cancel fractions, as otherwise you may lose marks.

The sum of probabilities

Suppose the probability that it rains tomorrow is $\frac{5}{8}$, then the probability that it does not rain tomorrow is $\frac{3}{8}$. When only one result can happen at a time the probabilities add up to 1:

$$P(\text{event occurs}) + P(\text{event does not occur}) = 1$$

If there are three possible probabilities, A, B and C, and only one can occur at once:

$$P(A) + P(B) + P(C) = 1$$

WORKIT!

A box contains coloured balls that are red, blue or green.

The table shows the number of balls of each colour.

	Red	Blue	Green
Number of balls	$2x + 1$	7	$x + 2$

A ball is chosen at random. The probability of the ball being blue is $\frac{7}{25}$.

Calculate the probability of choosing a red ball.

There are only red, blue and green balls in the bag, so the total probability must add up to 1.

$$P(\text{blue}) = \frac{\text{number of blue balls}}{\text{total number of balls}} = \frac{7}{25}$$

So total number of balls = 25

$2x + 1 + 7 + x + 2 = 25$ ◄——— Add up the number of balls to create an equation.

$3x + 10 = 25$

$3x = 15$

$x = 5$

Number of red balls = $2x + 1 = 2 \times 5 + 1 = 11$ ◄——— Use the calculated value of x to work out the number of red balls.

Probability of choosing a red ball = $\frac{11}{25}$

Sample space

In order to work out probability you need to be able to work out **outcomes**. The outcomes are the different possible combinations of what could happen. For example, when you toss two coins the possible outcomes are (HH, HT, TH and TT). You can use a **sample space** to show all the possible outcomes.

WORKIT!

A fair dice is rolled and a fair triangular spinner is spun. The scores are added together.

a Produce a sample space showing all the possible outcomes. ◄——— The outcomes are all the different ways of throwing the dice and spinning the spinner.

Spinner

		1	2	3
	1	2	3	4
	2	3	4	5
Dice	3	4	5	6
	4	5	6	7
	5	6	7	8
	6	7	8	9

The first column shows the score on the dice and the first row the score on the spinner.

The total scores are then filled in.

b Write down the total number of outcomes.

Total number of outcomes = 18 ◄——— The number of outcomes is the number of totals in the sample space diagram.

c Find the probability of obtaining a total score of 7.

$$P(7) = \frac{\text{number of ways of scoring 7}}{\text{total number of outcomes}} = \frac{3}{18} = \frac{1}{6}$$

◄——— The dice and spinner are both fair, so all the outcomes are equally likely.

NAILIT!

An event is an outcome or group of outcomes you are interested in, such as 'it rains tomorrow' or 'it rains on Saturday and is fine on Sunday'.

You can work out the total number of outcomes using the formula:

Total number of outcomes = number of ways each action can be carried out, multiplied together ◄———

For the example above, total number of outcomes = number of dice outcomes × number of spinner outcomes: 6 × 3 = 18.

WORKIT!

A restaurant has a fixed price menu offering a choice of 4 starters, 6 mains and 5 desserts.

a How many different meals are possible? Assume that all the meals consist of one starter, one main and one dessert.

Total number of meals = 4 × 6 × 5 = 120

b What is the probability of choosing the first item on each of the starters, mains and desserts menus?

P(1st starter, 1st main, 1st dessert) = $\frac{1}{120}$

> Each outcome (combination of starter, main and dessert) is equally likely.

WORKIT!

Four fair dice are rolled.

a Calculate the number of possible outcomes.

Total number of outcomes = number of ways each action can be carried out, multiplied together

= 6 × 6 × 6 × 6 = 1296

b Calculate the number of possible ways of all the dice showing prime numbers.

> The possible prime numbers are 2, 3 and 5.

Number of prime numbers on 1 dice = 3

Number of possible outcomes for all prime numbers = 3 × 3 × 3 × 3 = 81

c Calculate the probability of all the dice showing prime numbers.

Probability of all prime numbers = $\frac{81}{1296} = \frac{1}{16}$

DOIT!

Two coins are tossed. Work out the probability of one coin landing heads and the other landing tails.

CHECKIT!

1 A five-sided spinner is spun 3 times. Find the probability of the spinner showing an even number every time.

2 There are 20 counters in a bag. 4 are black, 7 are red and 9 are green. One counter is picked at random from the bag.

a Write down the probability of obtaining a white counter.

b Calculate the probability that the counter picked is black, giving your answer as a fraction in its simplest form.

c Calculate the probability that the counter picked is **not** green.

Probability experiments

A probability experiment is an experiment, trial or observation that can be repeated numerous times under the same conditions. Each outcome of a probability experiment must not be affected by any other outcome and cannot be predicted with certainty.

Examples of probability experiments include:

- tossing a coin: two possible outcomes, heads or tails
- rolling a six-sided dice: six possible outcomes, the numbers 1 to 6
- selecting a numbered ball (1–50) from a bag: 50 possible outcomes.

Relative frequency

Relative frequency is an estimate of probability. For example, suppose you rolled a fair (i.e. unbiased) dice 500 times and recorded your results.

Score	Frequency
1	91
2	77
3	99
4	74
5	93
6	66

$$\text{Relative frequency} = \frac{\text{frequency of event}}{\text{total frequency}}$$

Total frequency (see table) $= 91 + 77 + 99 + 74 + 93 + 66 = 500$.

The relative frequency for a score of $3 = \frac{99}{500} = 0.198$.

The relative frequencies can be used to create a table similar to the results table.

Score	1	2	3	4	5	6
Relative frequency	$\frac{91}{500} = 0.182$	$\frac{77}{500} = 0.154$	$\frac{99}{500} = 0.198$	$\frac{74}{500} = 0.148$	$\frac{93}{500} = 0.186$	$\frac{66}{500} = 0.132$

Relative frequency and probability

The relative frequency can be used to give an estimate of the probability.

The theoretical probability of a particular score on an unbiased dice $= \frac{1}{6}$ or 0.1666... as a decimal.

The relative frequencies are different from this value.

The more times the experiment is conducted, the closer the relative frequencies will be to the theoretical probability $\left(\frac{1}{6}\right.$ in this case$\left.\right)$.

> Note that all these relative frequencies add up to 1.

> The relative frequency will approach the theoretical frequency only over a very large number of trials.

WORKIT!

A five-sided spinner with sides numbered from 1 to 5 has the following relative frequencies of scores.

Score	1	2	3	4	5
Relative	x	0.05	$2x$	0.15	0.20

Calculate the value of x.

$$x + 0.05 + 2x + 0.15 + 0.20 = 1$$
$$3x + 0.4 = 1$$
$$3x = 0.6$$
$$x = 0.2$$

> The total relative frequencies have to add up to 1.

NAILIT!

Always check that you have taken into account all the possibilities before deciding that the probabilities add up to 1.

Expected frequency

To work out an estimate of the number of times an event occurs (**expected frequency**) we can use the following formula:

Expected frequency = probability of the event × number of events

WORKIT!

A seed picked at random from a large batch of seeds has a probability of growing into a plant with yellow flowers of 0.12.

There are 200 seeds in a packet. Work out an estimate for how many seeds will grow into plants with yellow flowers.

Expected number of yellow plants = probability of yellow plant
$$\times \text{ number of seeds}$$
$$= 0.12 \times 200$$
$$= 24$$

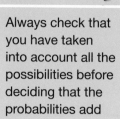

DOIT!

Create a revision card for relative and theoretical probability.

CHECKIT!

1 A six-sided spinner with numbers from 1 to 6 was spun 120 times. The results are shown in the table.

Number on spinner	1	2	3	4	5	6
Frequency	18	20	25	21	19	17

  a Work out the relative frequency for a score of 3, giving your answer as a decimal to 2 decimal places.

  b Work out the relative frequency for a score of 6, giving your answer as a decimal to 2 decimal places.

  c Sean says that the answers to parts a and b should be the same so the spinner is not fair (i.e. it is biased). Is Sean right? Explain your answer.

2 A machine fills cans with 330 ml of cola. Out of a batch of 500 cans, 20 of them contained less than 330 ml of cola.

  a Estimate the probability that the next can filled by the machine contains less than 330 ml of cola.

  b The machine fills 15 000 cans per day. Estimate how many would contain less than 330 ml of cola.

3 In a batch of apples it was found that the probability of an apple being bad was $\frac{3}{40}$.

In a similar batch of 600 apples, how many would be expected to be bad?

The AND and OR rules

If events A and B are **mutually exclusive**, it means that event A can happen or event B can happen but they cannot both happen at the same time. When an event has no effect on another event, they are said to be **independent events**.

NAILIT!

Note the difference between independent events and mutually exclusive events.

There are two important laws that need to be remembered when working out probabilities. They are the **addition rule** and the **multiplication rule**. The addition rule is sometimes called the **OR rule** and the multiplication law is sometimes called the **AND rule**.

SNAPIT!   AND and OR rules

You must learn these formulae.

Probability of A or B happening: $P(A \text{ OR } B) = P(A) + P(B)$ ◄

> This formula only applies to mutually exclusive events.

Probability of A and B happening: $P(A \text{ AND } B) = P(A) \times P(B)$ ◄

> This formula only applies to independent events.

WORKIT!

1 A bag contains 3 red balls and 7 black balls. Ahmed picks one at random and notes the colour. He puts it back into the bag and picks another ball at random.

What is the probability that both balls are red?

$P(\text{1st red}) = \frac{3}{10}$ and $P(\text{2nd red}) = \frac{3}{10}$

$P(\text{red AND red}) = P(\text{1st red}) \times P(\text{2nd red}) = \frac{3}{10} \times \frac{3}{10} = \frac{9}{100}$ ◄

> The two events (picking each ball) are independent, as the balls in the bag are the same for each pick.

2 A computer manufacturer submits bids for three projects, A, B and C.

The probabilities of getting the order for projects A, B and C are $\frac{2}{5}$, $\frac{1}{3}$ and $\frac{1}{4}$, respectively. The probability of getting the order for each project is independent of the other projects.

Calculate the probability of the following events, making your methods clear. Give each answer as a fraction in its simplest form.

a The company gets all three orders.

$$P(A \text{ and } B \text{ and } C) = \frac{2}{5} \times \frac{1}{3} \times \frac{1}{4} = \frac{1}{30}$$ ← Use the AND rule.

b The company gets exactly one order.

A	Not B	Not C	$\frac{2}{5} \times \frac{2}{3} \times \frac{3}{4} = \frac{12}{60}$
Not A	B	Not C	$\frac{3}{5} \times \frac{1}{3} \times \frac{3}{4} = \frac{9}{60}$
Not A	Not B	C	$\frac{3}{5} \times \frac{2}{3} \times \frac{1}{4} = \frac{6}{60}$

$$P(\text{exactly one order}) = \frac{12}{60} + \frac{9}{60} + \frac{6}{60} = \frac{27}{60} = \frac{9}{20}$$

Make a table showing the possible outcomes for winning exactly one project.

P(not getting an order) = 1 − P(getting an order)

Use the AND rule to work out the probability of each of these outcomes.

DO IT!

Make a revision poster for the AND rule and the OR rule, including the necessary conditions.

✓ CHECKIT!

1 James passes two sets of traffic lights on his way to work. The probability he is stopped by the first set of traffic lights is 0.2 and the probability he is stopped by the second set of lights is 0.3. The two sets of traffic lights are independent.

  a Explain in relation to this question what independent events are.

  b Work out the probability that James is stopped by both sets of lights.

  c Work out the probability that James is not stopped by either set of lights.

2 The probability that Nala takes a packed lunch to school is $\frac{1}{4}$.

The probability that Nala cycles to school is $\frac{2}{3}$.

The probability that Nala has remembered her homework is $\frac{7}{8}$.

  a Work out the probability all of these events take place.

  b Work out the probability that none of the events takes place.

Tree diagrams

The probabilities of two or more events can be shown on a **tree diagram**.

Independent events

> The two events (picking the balls out) are independent because the first ball is put back into the bag.

WORKIT!

A bag contains 5 red and 4 blue balls. A ball is picked at random from the bag, the colour is noted and the ball is put back into the bag. Another ball is picked at random and its colour noted.

a Draw a tree diagram to show these events.

b Work out the probability that both balls are red.

$$P(\text{red and red}) = \frac{5}{9} \times \frac{5}{9} = \frac{25}{81}$$

c Work out the probability that the balls are of different colours.

$$P(\text{red then blue}) = \frac{5}{9} \times \frac{4}{9} = \frac{20}{81}$$

$$P(\text{blue then red}) = \frac{4}{9} \times \frac{5}{9} = \frac{20}{81}$$

$$P(\text{different colours}) = \frac{20}{81} + \frac{20}{81} = \frac{40}{81}$$

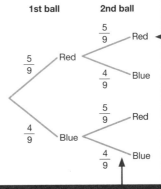

> Because you can have either of these, use the OR rule.

> As the balls are replaced, the probability for the second ball is the same as for the first ball. The probabilities at each branch add up to 1 as you are certain to pick either a red or a blue ball.

> To work out the probability for a particular combination (e.g. a red and a red), multiply the probabilities on the two branches.

Dependent events

In **dependent** events, one event influences another. This tree diagram shows the same problem as above, except that this time the first ball is **not** replaced. The probability for the second ball depends on what colour was picked for the first ball.

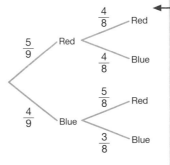

> If a red ball is picked and not replaced, there are only 4 red balls and 8 balls in total for the second pick.

175

WORKIT!

The probability that it rains on Monday is 0.3.

If it rained on Monday, the probability of it not raining on Tuesday is 0.4.

If it did not rain on Monday, the probability of it not raining on Tuesday is 0.6.

a Draw a probability tree diagram showing this information.

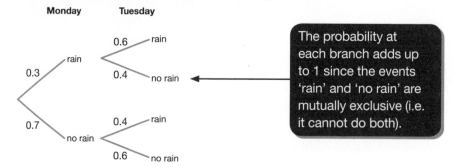

Monday   **Tuesday**

0.3 — rain

0.7 — no rain

0.6 — rain

0.4 — no rain

0.4 — rain

0.6 — no rain

> The probability at each branch adds up to 1 since the events 'rain' and 'no rain' are mutually exclusive (i.e. it cannot do both).

b It rained on only one of the two days. Is it more likely that it rained on Monday or Tuesday? Show your working.

P(only rained on Monday) = P(rain Monday) × P(no rain Tuesday)

= 0.3 × 0.4 = 0.12

P(only rained on Tuesday) = P(no rain Monday) × P(rain Tuesday)

= 0.7 × 0.4 = 0.28

It is more likely that it rained on Tuesday.

NAILIT!

Make sure you state on which day it is more likely to have rained. It is not up to the examiner to decide this from your working out.

You may not always be told to draw the tree diagram. You may also need to use algebra to work out the answer.

> Draw a tree diagram.

WORKIT!

A box contains cartons of orange juice and apple juice in the ratio 2 : 1.
Two cartons of juice are taken out of the box at random.
The probability that both cartons are apple juice is $\frac{7}{69}$.

How many cartons of apple juice are there in the box?

Let the number of apple juice cartons be x. ◄ Use a variable for one part of the ratio.

$P(2 \text{ apple}) = \frac{x}{3x} \times \frac{x-1}{3x-1} = \frac{7}{69}$

> $3x$ is total number of cartons using both parts of the ratio

$69x(x-1) = 7 \times 3x(3x-1)$

$23(x-1) = 7(3x-1)$ ◄ Divide both sides by $3x$.

$23x - 23 = 21x - 7$

$2x = 16$

$x = 8$ ◄ Solve the equation for x.

There are 8 cartons of apple juice in the box.

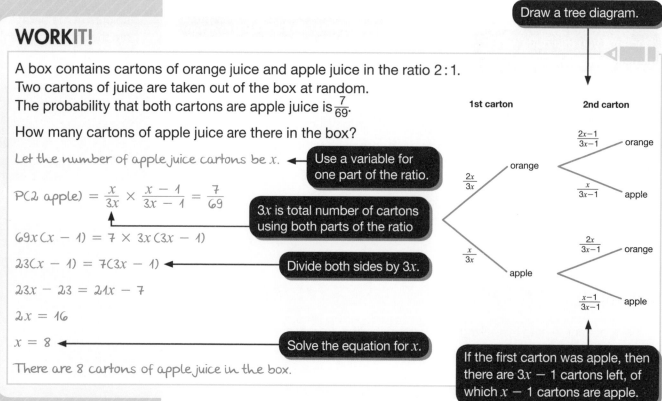

1st carton   **2nd carton**

$\frac{2x}{3x}$ orange

$\frac{x}{3x}$ apple

$\frac{2x-1}{3x-1}$ orange

$\frac{x}{3x-1}$ apple

$\frac{2x}{3x-1}$ orange

$\frac{x-1}{3x-1}$ apple

> If the first carton was apple, then there are $3x - 1$ cartons left, of which $x - 1$ cartons are apple.

Frequency trees

The results of a probability experiment can be displayed using a **frequency tree**. The frequency tree can then be used to estimate the probability of certain scenarios.

WORKIT!

There are 350 staff working in an office. The ratio of males to females is 4:3.

For the male staff, the ratio of the number of full-time staff to the number of part-time staff is 3:1.

For the female staff the ratio of the number of part-time staff to the number of full-time staff is 2:1.

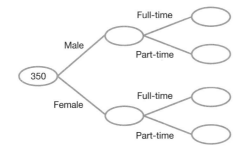

a Complete the frequency tree.

Males : females = 4 : 3   7 parts = 350, so 1 part = 50

Number of males = 4 × 50 = 200

Number of females = 3 × 50 = 150

Males: full-time : part-time = 3 : 1   4 parts = 200, so 1 part = 50

Number of males full-time = 3 × 50 = 150

Number of males part-time = 1 × 50 = 50

Females: part-time : full-time = 2 : 1   3 parts = 150, so 1 part = 50

Number of females part-time = 2 × 50 = 100

Number of females full-time = 1 × 50 = 50

> Use the ratios to work out the numbers of males and females and the numbers of full- and part-time staff.

> Fill in the frequency tree with the calculated values.

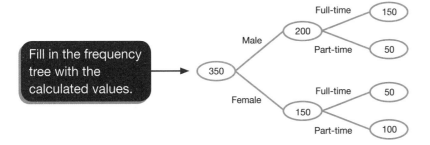

> **NAIL**IT!
>
> Read the question carefully. Here, full-time and part-time are mentioned in a different order on the second occasion.

There are 150 female staff out of a total of 350 staff.

Out of 150 female staff, 50 of them work full-time.

b Using the frequency tree, work out the probability that a member of staff chosen at random is a female who works full-time.

Probability a female is chosen $= \frac{150}{350} = \frac{3}{7}$

Probability that the female works full-time $= \frac{50}{150} = \frac{1}{3}$

P(female who works full-time) $= \frac{3}{7} \times \frac{1}{3} = \frac{1}{7}$ ◄ Use the AND rule.

 **DOIT!**

Write some revision cards containing all the main points and formulae for probability.

✓ CHECKIT!

1 There are 3 red and 7 blue counters in a bag. Two counters are picked at random, one at a time, from the bag.

Work out the probability that:

a two red counters are chosen

b a red and a blue counter are chosen.

2 Hannah puts green and blue balls into an empty bag in the ratio 2:3.

Zak takes at random two balls from the bag. The probability he takes two blue balls is $\frac{33}{95}$.

How many balls did Hannah put in the bag?

3 Saskia goes to the airport to catch a flight. The probability that there is a long queue to check in is 0.7. The probability that there is a long queue at security is 0.5.

a Copy and complete the tree diagram.

b Calculate the probability that there is a short queue at both check-in and security.

c Calculate the probability of a long queue at either check-in or security, or at both.

Check in

Security check

0.7 — long queue

0.5 — long queue

...... — short queue

...... — short queue

...... — long queue

...... — short queue

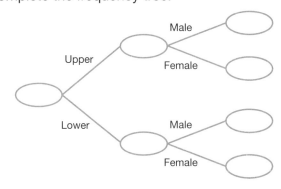

4 A secondary school has an upper school of 450 students and a lower school of 500 students. In the upper school, 60% of the students are male and in the lower school 55% are male.

a Copy and complete the frequency tree.

Upper

Male

Female

Lower

Male

Female

b A student is chosen at random from the school. Find the probability that the student is male.

Venn diagrams and probability

Venn diagrams can be used to work out probability.

The **universal set** is the whole collection of things, called **elements**, being considered. Usually an element is a number (e.g. {3, 5, 8} or an item {Jack, Queen, King}).

Sets are parts of the universal set and are shown as labelled circles/ovals on the Venn diagram. Sets contain some of the elements in the universal set and no other elements.

The universal set is labelled with the Greek letter ξ (ξ, pronounced ksi).

Curly brackets are used to show sets.

The Venn diagram shows the universal set ξ = {1, 2, 3, 4, 5, 6, 7, 8, 9, 10}, set A = {2, 3, 5, 6, 7, 8} and set B = {1, 2, 4, 6, 9}.

The number of elements in set A is referred to as n(A): in this example n(A) = 6.

SNAP IT! Venn diagrams

Here are the terms used for different areas on the Venn diagram:

Expression	Term	Meaning	Set of numbers
A ∩ B	A intersect B	All the elements in set A which are also in set B	{2, 6}
A ∪ B	A union B	All the elements in both set A and set B	{1, 2, 3, 4, 5, 6, 7, 8, 9}
A′	A complement	All the elements in the universal set that are not in set A	{1, 4, 9, 10}

WORKIT!

The Venn diagram shows all the factors of 36 and 40.

Set A is the set of all the factors of 36.

Set B is the set of all the factors of 40.

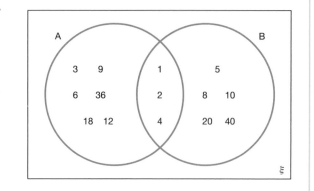

a Write down all the numbers in A ∩ B and describe their significance.

A ∩ B = {1, 2, 4}, all the numbers that are a factor of both 36 and 40

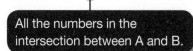

All the numbers in the intersection between A and B.

b A number is picked at random from the set A ∪ B. Find the probability that it is an even number.

$$P(\text{even number in A} \cup \text{B}) = \frac{10}{14} = \frac{5}{7}$$

> A ∪ B is the set of all the numbers in A and B: 14 numbers of which 10 are even.

c A number is picked at random from the set B′.
Find the probability that the number is a multiple of 9.

$$B' = \{3, 6, 9, 12, 18, 36\}$$

$$P(\text{multiple of 9 in B}') = \frac{3}{6} = \frac{1}{2}$$

> Three elements of B′ are multiples of 9.

NAILIT!

When listing numbers in the complement, don't forget the numbers that are outside both A and B.

WORKIT!

There are 43 members in a youth club. 25 play badminton, 16 play chess and 20 play tennis. 5 play all three games. 2 play chess and tennis but not badminton. 7 play tennis and badminton only. 6 members play only chess. 4 members do not play any games.

a Draw a Venn diagram showing the above information.

16 play chess, so number who play badminton and chess only
= 16 − (6 + 5 + 2) = 3

25 play badminton, so the number who play badminton only
= 25 − (3 + 5 + 7) = 10

20 play tennis, so the number who play tennis only = 20 − (7 + 5 + 2) = 6

> Use the information in the question to work out the other values for the Venn diagram.

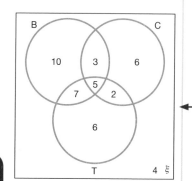

> First add all the easy information to the Venn diagram.

> Put the values into the diagram as you work them out.

b A member of the club is picked at random. Find the probability that this member only plays badminton.

$$P(\text{badminton only}) = \frac{10}{43}$$

c Out of those members who play badminton, find the probability that they play only **one** other game.

P(badminton and only one other game)

$$= \frac{7 + 3}{10 + 7 + 3 + 5} = \frac{10}{25} = \frac{2}{5}$$

NAILIT!

Check that all the numbers on the Venn diagram add up to the total given in the question.

DO IT!

Produce pairs of cards containing a shaded area on a Venn diagram and the expression for the area (e.g. A ∩ B). Practise matching the cards.

CHECK IT!

1 The diagram below is a Venn diagram.

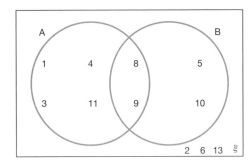

a Write down the numbers that are in the set:

i A ∪ B   **ii** A ∩ B   **iii** A′

b One of the numbers is picked at random. Find P(B′).

2 In a sixth form of 180 students:
84 students take at least one science
40 study all three sciences
1 student studies only physics
12 study physics and biology only
48 study physics and chemistry
15 study biology and chemistry, but not physics
65 study chemistry.

a A student is chosen at random from those who take science. What is the probability that the student takes all three sciences?

b A student is chosen at random from those who take science. What is the probability that the student takes only one science?

c Out of those students studying physics, find the probability that they also study chemistry.

3 The universal set is the set of integers from 1 to 20. Set A is the set of square numbers. Set B is the set of all even numbers.

a Complete the following Venn diagram.

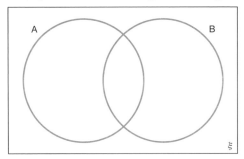

b An integer is chosen at random. Find the probability that it is not an even or square number.

Probability

1 A bag contains 4 red counters, 3 green counters and 3 white counters.

A counter is taken at random from the bag and not replaced. Then a second counter is taken from the bag.

a Complete the tree diagram.

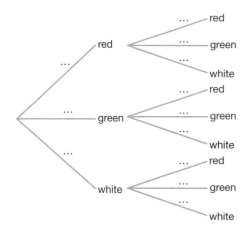

b Calculate the probability that the two counters taken from the bag are the same colour.

c Calculate the probability that the two counters taken from the bag are different colours.

2 A bag contains 3 red balls and x black balls.

Two balls are removed at random from the bag.

The probability that two black balls are chosen is $\frac{7}{15}$.

a Show that $4x^2 - 25x - 21 = 0$.

b Find the total number of balls in the bag.

c Find the probability of the two balls being different colours.

3 60 students were asked which sports they liked watching from football, tennis and motor racing.

All 60 students liked watching at least one sport.

10 students liked watching all three sports.

3 liked watching football and motor racing only.

15 liked watching tennis and motor racing.

29 liked watching motor racing.

18 liked watching football and tennis.

41 liked watching football.

a What is the probability that a student chosen at random from the group only liked to watch motor racing?

b A student is chosen from those who liked motor racing. What is the probability that the student also liked tennis?

4 In a certain country, out of 500 defendants being tried by the law courts 75% committed the crime. Of those who committed the crime, 80% were found guilty. Of those who did not commit the crime, 20% were found guilty.

a Complete the frequency tree to show this information.

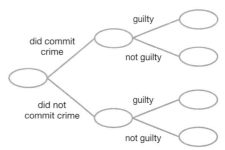

b A defendant was chosen at random from the 500 defendants. Find the probability that the defendant was found guilty.

c A defendant was chosen at random from those who did not commit the crime. Find the probability that the defendant was found guilty.

Statistics

Sampling

It is not usually possible to carry out a survey or a poll on an entire **population** (e.g. people eligible to vote in an election, students in a school, people who attend a sporting event). Instead a smaller number of people called a **sample** are surveyed.

The sample should be large enough to represent the population (e.g. a sample of 50 out of a population of 10 000 would be too small). The larger the sample size, the more accurate the results. The sample must also be **representative** of the population (e.g. same proportion of males/females and age groups).

Random sampling

The sample should be taken randomly, with each member of the population having an equal chance of being chosen. To ensure a **random** sample, you could:

- number each person and then use the random select on a calculator or computer
- write names on paper, fold them, jumble them up in a container and then pick them out.

Bias

A **biased** sample will not reflect the population accurately. The following should be considered when sampling:

- **The time the survey is done**: doing a survey at 8.30 am outside a train station on a weekday is less likely to include people who do not work.
- **Where the survey is done**: trying to find out the number of holidays people take by interviewing people as they leave a travel agent shop would bias the results towards people who take holidays.
- **The gender of people in the survey**: the sample should have the same percentage of males/females as the population.
- **The age of the people in the survey**: the sample should have the same age composition as the population.

Stratified samples

Stratified samples ensure that the sample mirrors the make-up of the population.

WORKIT!

There are 208 employees in a company. The table shows numbers of employees in each department.

A survey is to be conducted with a sample of 50 employees.

Sales	Administration	Manufacturing	Delivery	Accounts	Management
56	33	76	21	12	10

a   Explain the difference between a population and a sample, illustrating your answer with the information above.

   The population is the whole group, in this case all the employees.

   The sample is a smaller group chosen from the population. In this case, it is the 50 employees to be used for the survey.

   A stratified sample of 50 employees is needed for the survey.

b   Calculate the number of employees in the sales department that should be chosen.

   % of the employees who work in sales $= \dfrac{\text{number who work in sales}}{\text{total of all employees}} \times 100$

   > Find this percentage of the number in the sample.

   $= \dfrac{56}{208} \times 100$

   $= 26.9\%$

   Number of sales employees in sample $= 26.9\%$ of $50 = 0.269 \times 50 = 13.45$

   The sample will include 13 members of the sales department.

   > Round the calculated value to the nearest whole number.

NAILIT!

An even quicker way of doing this would be to multiply 50 by the fraction $\dfrac{56}{208}$.

DOIT!

Think about how you could perform a survey on students in your school. How many students should be in the sample? How would you stratify the sample?

CHECKIT!

1   A youth club has 350 members. The club leader wants to survey the members to check the club provides the sorts of activities young people like. He has decided to conduct a survey of 50 members selected at random.

The table shows the number of male and female members.

Male	Female
185	165

Work out the number of males that should be chosen for the sample.

2   The table gives information about the numbers of students in a school for each of the two GCSE years.

	Male	Female
Year 10	146	164
Year 11	155	175

A stratified sample of 50 students is to be taken for a survey stratified by year and by gender.

Work out the number of male students in year 10 that should be in this sample.

Two-way tables, pie charts and stem-and-leaf diagrams

Two-way tables

Two-way tables are used to show information in two categories. The following example shows data on gender and favourite food.

WORKIT!

The two-way table gives information about the favourite food, chosen from Indian, Chinese and Italian, of 261 students in a school.

	Indian	Chinese	Italian	Total
Boys	75		44	147
Girls		38		114
Total	125	66		

a Complete the two-way table.

	Indian	Chinese	Italian	Total
Boys	75	28	44	147
Girls	50	38	26	114
Total	125	66	70	261

The numbers must add up across the rows and down the columns to give the totals.

Start with a column or row that is only missing one value: boys liking Chinese = 147 − 75 − 44 = 28.

b A student is picked at random from the school. What is the probability that the student's favourite food is Chinese? Give your answer as a fraction in its simplest form.

$P(Chinese) = \frac{66}{261} = \frac{22}{87}$

c A boy is picked at random. Find the probability that his favourite food is Indian or Italian.

$P(boy's\ favourite\ food\ is\ Indian\ or\ Italian) = \frac{75 + 44}{147} = \frac{119}{147} = \frac{17}{21}$

Tabulating discrete and continuous data

Numerical data can be **discrete** or **continuous**. Continuous data can take any value, for example mass and height. Discrete data can only have specific values, for example the number of people in a queue.

Age of cars (nearest whole year)	Tally	Frequency
0–2		
3–5		
6–8		
9–11		
12 or more		

The data is discrete because the age of the cars is given to the nearest whole number.

Suppose you did a survey of the ages of cars in a car park. You could produce a list of all the ages but this would be cumbersome, particularly if there is a lot of data. Instead, put the ages in whole years into **classes** (groups) in a table.

With continuous data, the classes are shown using inequalities. Here are the heights of 75 dogs taking part in a dog show.

Height of dog (h cm)	Frequency
$0 < h \le 20$	6
$20 < h \le 40$	25
$40 < h \le 60$	28
$60 < h \le 80$	10
$80 < h \le 100$	6

Pie charts

A **pie chart** shows the division of a total into its separate parts. To work out the angle for each part:

1. Find the total frequency.

2. Divide 360° by the total frequency to find the number of degrees represented by one item.

3. Multiply this by each of the frequencies to find the angle for each category.

4. Use these angles to draw the pie chart.

> $80 < h \le 100$ is a class. There are five classes in this table.

WORKIT!

The table gives information about the manufacturers of cars in a car park.

Manufacturer	Frequency
Ford	12
Vauxhall	18
Nissan	5
Toyota	10
Mercedes	15

Draw an accurate pie chart showing this information.

$12 + 18 + 5 + 10 + 15 = 60$ ◄── Add up the frequencies.

60 cars $= 360°$ so angle representing 1 car $= \frac{360}{60} = 6°$ ◄── Multiply the frequencies by this angle to find the angle for each manufacturer.

Manufacturer	Frequency	Angle
Ford	12	$12 \times 6 = 72°$
Vauxhall	18	$18 \times 6 = 108°$
Nissan	5	$5 \times 6 = 30°$
Toyota	10	$10 \times 6 = 60°$
Mercedes	15	$15 \times 6 = 90°$

Always check that the angles add up to 360°.

NAIL IT!

Always make sure that classes do not have values that overlap. For example, with two classes $0 < h \le 20$ and $20 \le h \le 40$, a height of 20 cm would be in both classes.

Use compasses, a ruler and a protractor to draw the pie chart.

Stem-and-leaf diagrams

Stem-and-leaf diagrams are used to show a small number of data items.

WORKIT!

Sara is performing a survey on how many music downloads students in her class make each month.

Here are her results.

12	19	8	61	45	0	8	41	46	51
18	23	20	15	39	31	21	20	18	3
7	53	60	36	15					

Draw an ordered stem-and-leaf diagram of Sara's results.

The column on the left is used for the tens. 0 represents no tens, 1 represents one ten and so on.

0	0 3 7 8 8
1	2 5 5 8 8 9
2	0 0 1 3
3	1 6 9
4	1 5 6
5	1 3
6	0 1

Key: 3|1 represents 31 downloads

The numbers to the right of the vertical line represent the units. All the numbers are in ascending order.

NAILIT!

Always include a key.

NAILIT!

Always check that the number of data values in the stem-and-leaf diagram is the same as the number of values in the question.

✓ CHECKIT!

1 The table gives information about newt species in a pond.

Species	Frequency
Smooth newt	13
Great crested newt	7
Palmate newt	10

a Draw a pie chart to show this information.

b On a pie chart for another pond the angle sector for the smooth newt was 135°. Does this mean that fewer smooth newts were in the other pond? Explain your answer.

2 Here is a list of times it takes in minutes to drive into town on different days.

22	30	35	30
23	31	42	45
33	36	47	40
23	35	24	41

Draw an ordered stem-and-leaf diagram to show this information.

Line graphs for time series data

Some quantities (e.g. sales of fireworks or barbecues, temperature, rainfall) vary in a repeating pattern over time. When these quantities are plotted against time they show repeating peaks and troughs, making it difficult to spot **trends** (e.g. increasing or decreasing values over time).

> These events are affected by the seasons.

Using a **moving average** makes it easier to spot trends. This smooths out the peaks and troughs so that the graph is straighter.

Moving averages

Suppose you have the following data.

> Q1 2016 means the first quarter (January to March) of 2016.

	Q1 2015	Q2 2015	Q3 2015	Q4 2015	Q1 2016	Q2 2016	Q3 2016	Q4 2016
Sales (thousands)	7	12	6.5	4	6	11	5.5	3.5

The graph of this data shows a lot of variation and it is hard to see the trend.

To find the 4-point moving average (to 1 decimal place):

1 Find the mean of the first four values:

$$\frac{7 + 12 + 6.5 + 4}{4} = 7.4$$

Plot this value at the centre of the period covered by the four values (i.e. midway between Q2 2015 and Q3 2015).

2 Move one unit of time along (in this case, Q2 2015) and find the mean of the next four values:

$$\frac{12 + 6.5 + 4 + 6}{4} = 7.1 \text{ plotted midway between Q3 and Q4 2015.}$$

3 Repeat step 2 until the end of the data:

$$\frac{6.5 + 4 + 6 + 11}{4} = 6.9 \text{ plotted midway between Q4 2015 and Q1 2016}$$

$$\frac{4 + 6 + 11 + 5.5}{4} = 6.6 \text{ plotted}$$
midway between Q1 and Q2 2016

$$\frac{6 + 11 + 5.5 + 3.5}{4} = 6.5 \text{ plotted}$$
midway between Q2 and Q3 2016.

Plotting the original data and the 4-point moving average gives this graph.

The trend line shows that there is a slight general downward trend in sales.

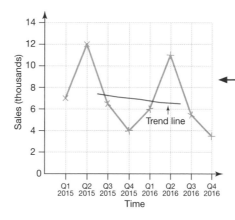

> With a 2- or 3-point moving average you would find the average of 2 or 3 values, respectively. This would be less effective in smoothing out seasonal variations when you are using quarterly figures.

WORKIT!

A garden centre sells fireworks. Here is a table showing the number of boxes of fireworks they sell over six months.

Month	September	October	November	December	January	February
Sales	31	29	60	49	35	63

a Work out the 3-point moving averages. The first two have been worked out for you:

40, 46, ..., ...

> Starting from the November figure, find the average over three months.

Third 3-point moving average $= \dfrac{60 + 49 + 35}{3} = \dfrac{144}{3} = 48$

> If you were plotting the graph, this would be plotted at January.

Fourth 3-point moving average $= \dfrac{49 + 35 + 63}{3} = \dfrac{147}{3} = 49$

b Use your answer to part a to describe the trend in the sales of fireworks over the months shown.

The figures show increasing fireworks sales.

> The values are all increasing, so the trend is increasing sales.

DOIT!

Draw a flowchart to work out a 4-point moving average for a set of figures over a 12-month period.

 CHECKIT!

1 The sales of a maths textbook for the first six months of the year are recorded here.

Month	Jan	Feb	Mar	Apr	May	June
Sales	936	939	903	870	882	810

a Work out the 3-point moving averages.

b Describe the trend in sales over the period for the data.

2 This table records the percentage of people who live in a village visiting their local shop.

Year	1996	2000	2004	2008	2012	2016
Percentage	70	65	57	53	51	49

a Plot the time series graph. Make the x-axis go up to at least 2025.

b Describe the trend in the percentage of people using the local shop over the last 25 years.

c If the trend continues, what percentage of the people are likely to be using the village shop in 2025?

d Explain why your prediction in part c may not be reliable.

Averages and spread

In statistics there are three **averages**: mean, mode and median.

An average is a typical value that is representative of a whole set of data.

The **mean** of a set of n items is the sum of the items divided by n:

$$\text{Mean} = \frac{\text{total of the items}}{\text{number of items } (n)}$$

To work out the **median** of a set of numbers:

1. Arrange the numbers in ascending order (i.e. smallest first).

2. Find the middle number; if there are two middle numbers, find the mean of them.

The **mode** is the most common value (the value with the greatest frequency).

WORKIT!

Here is a set of 14 values:

7, −8, 0, 4, −1, −2, 3, 2, 3, 2, −3, 2, 1, 1

Find the mean, median and mode.

$$\text{Mean} = \frac{7 + (-8) + 0 + 4 + (-1) + (-2) + 3 + 2 + 3 + 2 + (-3) + 2 + 1 + 1}{14}$$

$$= \frac{11}{14}$$

$$= 0.79 \text{ (to 2 d.p.)}$$

Arranging the values in ascending order:

−8, −3, −2, −1, 0, 1, 1, 2, 2, 2, 3, 3, 4, 7

There is an even number of values: the middle values are the 7th and 8th ones.

$$\text{Median} = \frac{1 + 2}{2} = 1.5$$

$$\text{Mode} = 2$$

2 is the most frequent value in the set, as it appears three times.

NAILIT!

The **median** is the halfway value. The **lower quartile** is the boundary between the first and second quarters of data. The **upper quartile** is the boundary between the third and fourth quarters of data. These are worked out in a similar way to the median.

NAILIT!

Always check the number of values after rearranging them to make sure you have included all of them.

WORKIT!

The mean age of four members in a group is 22. One new member joins the group and the mean age falls to 21. How old is the new member?

$$\text{Mean age of four members} = 22 = \frac{\text{total age}}{4}$$

$$\text{Total age} = 22 \times 4 = 88$$

Let the age of the member joining the group $= x$

$$\text{Mean age of 5 members} = \frac{88 + x}{5}$$

$$21 = \frac{88 + x}{5}$$

$$105 = 88 + x$$

$$x = 17 \text{ years}$$

NAILIT!

It is easier to spot the mode after putting the numbers in order.

DOIT!

Write a revision card summarising how to find each average and the lower and upper quartiles.

Estimating the mean of a grouped distribution

When there is a large number of data items, it is usual to put them into classes (e.g. $20 < t \leq 30$). You can't work out an exact mean, as you don't have the individual data values. However, you can **estimate** the mean using the mid-interval value for each class.

WORKIT!

Sally recorded the times it took her to walk to school over 30 days.
The table gives information about her data.

Time (t minutes)	Frequency		
$20 < t \leq 30$	4		
$30 < t \leq 40$	8		
$40 < t \leq 50$	16		
$50 < t \leq 60$	2		

NAILIT!

The blank columns in this table indicate that there is some working out to do.

Work out Sally's mean time to walk to school in minutes, correct to 3 significant figures.

Time (t minutes)	Frequency	Mid-interval value	Frequency × mid-interval value
$20 < t \leq 30$	4	25	4 × 25 = 100
$30 < t \leq 40$	8	35	8 × 35 = 280
$40 < t \leq 50$	16	45	16 × 45 = 720
$50 < t \leq 60$	2	55	2 × 55 = 110

Create two new columns: the mid-interval value and frequency × mid-interval value. The mid-interval value is the midpoint of the range.

$$\text{Estimated mean} = \frac{100 + 280 + 720 + 110}{30} = 40.33 = 40.3 \text{ minutes (to 3 s.f.)}$$

Divide the total of all the frequency × mid-interval values by the total frequency.

CHECKIT!

1 The mean age of boys in a youth club is 13 and the mean age of the girls is 14. There are 10 boys and 12 girls. Calculate the mean age for the club members.

2 There are 25 students in a maths class. The 10 boys in the class got a mean mark of 50% in a maths exam. The 15 girls got a mean mark of 62% in the same exam.

Joshua says that the mean mark for the entire class can be worked out as

$$\frac{50 + 62}{2} = 56\%.$$

Explain why Joshua is wrong and work out the correct mean mark, showing your working.

3 A company keeps a record of its employees' taxi expenses over a month.

Cost (£C)	Frequency	Mid-interval value	Frequency × mid-interval value
$0 < C \leq 4$	12		
$4 < C \leq 8$	8		
$8 < C \leq 12$	10		
$12 < C \leq 16$	5		
$16 < C \leq 20$	2		

a Copy and complete the table.

b Calculate an estimate for the mean taxi fare for the month.

Histograms

Histograms consist of vertical bars, but they are different from bar charts. In a histogram it is the **areas** of the bars that represent the frequency, not the height.

Histograms:

- have numbers on both axes
- can have bars of different widths
- have frequency density rather than frequency on the vertical axis.

SNAP IT! Histograms

Frequency (area of bar) = **class width** (width of bar) × **frequency density** (the bar's height)

You often need to rearrange this formula to give:

$$\text{Frequency density} = \frac{\text{frequency}}{\text{class width}}$$

The class width is found by subtracting the higher value from the lower value of the class interval: for $5 < d \le 9$, class width = $9 - 5 = 4$.

WORK IT!

1 The table gives information about the diameter of ripe tomatoes in a greenhouse.

Diameter (d centimetres)	Frequency
$0 < d \le 2$	14
$2 < d \le 3$	16
$3 < d \le 5$	24
$5 < d \le 9$	30

Draw a histogram to show this information.

Draw the histogram with diameter (d cm) on the horizontal axis and frequency density on the vertical axis.

Diameter (d centimetres)	Frequency	Frequency density
$0 < d \le 2$	14	$\frac{14}{2} = 7$
$2 < d \le 3$	16	$\frac{16}{1} = 16$
$3 < d \le 5$	24	$\frac{24}{2} = 12$
$5 < d \le 9$	30	$\frac{30}{4} = 7.5$

Work out the frequency densities by dividing each frequency by its class width.

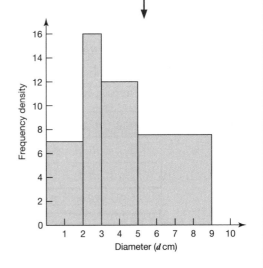

2 This histogram shows the ages of football team supporters travelling on a plane to a match.

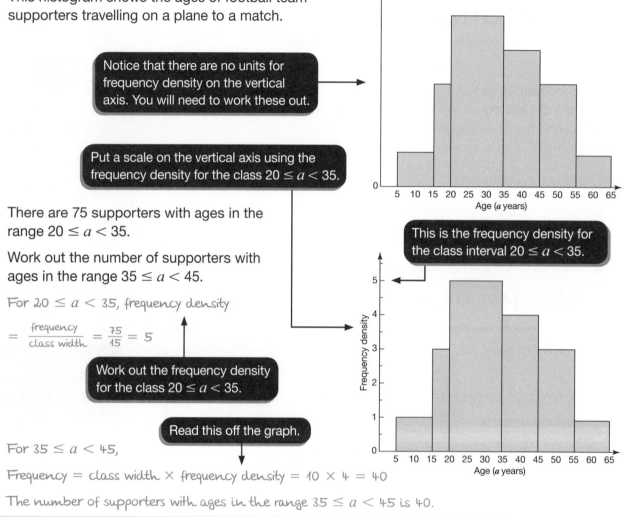

Notice that there are no units for frequency density on the vertical axis. You will need to work these out.

Put a scale on the vertical axis using the frequency density for the class $20 \leq a < 35$.

There are 75 supporters with ages in the range $20 \leq a < 35$.

This is the frequency density for the class interval $20 \leq a < 35$.

Work out the number of supporters with ages in the range $35 \leq a < 45$.

For $20 \leq a < 35$, frequency density

$= \dfrac{\text{frequency}}{\text{class width}} = \dfrac{75}{15} = 5$

Work out the frequency density for the class $20 \leq a < 35$.

Read this off the graph.

For $35 \leq a < 45$,

Frequency = class width × frequency density = $10 \times 4 = 40$

The number of supporters with ages in the range $35 \leq a < 45$ is 40.

DOIT!

Draw a flowchart for drawing a histogram.

CHECKIT!

1 The table and histogram show information about the masses of kittens.

Copy and complete the table and the histogram.

Mass (*m* pounds)	Frequency
$0.0 < m \leq 0.5$	30
$0.5 < m \leq 1.0$	
$1.0 < m \leq 1.5$	13
$1.5 < m \leq 2.0$	21
$2.0 < m \leq 2.5$	
$2.5 < m \leq 3.0$	38

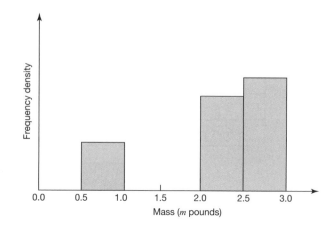

Cumulative frequency graphs

Cumulative frequency graphs have a running total of the frequency (called the **cumulative frequency**) on the vertical axis and the quantity you want information about on the horizontal axis.

You can find out the median and the lower and upper quartiles by reading them off the graph.

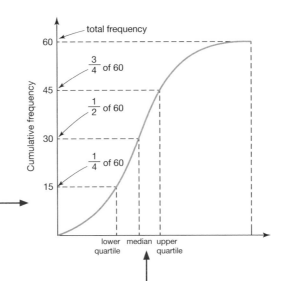

Interquartile range = upper quartile − lower quartile

WORKIT!

A perfume counter in a duty free store keeps records of the prices of bottles of perfume they sell in a day.

Cost (£C)	Frequency	Cumulative frequency
$0 < C \leq 20$	28	
$20 < C \leq 40$	42	
$40 < C \leq 60$	35	
$60 < C \leq 80$	45	
$80 < C \leq 100$	50	
$100 < C \leq 120$	15	

a Write down the modal class interval.

Modal class interval is $80 < C \leq 100$.

This is the class interval with the highest frequency.

b Copy and complete the cumulative frequency table.

Cost (£C)	Frequency	Cumulative frequency
$0 < C \leq 20$	28	28
$20 < C \leq 40$	42	70
$40 < C \leq 60$	35	105
$60 < C \leq 80$	45	150
$80 < C \leq 100$	50	200
$100 < C \leq 120$	15	215

$28 + 42 = 70$

The last entry in cumulative frequency is the total frequency. In this case it is the total number of bottles of perfume sold.

195

c Draw the cumulative frequency diagram.

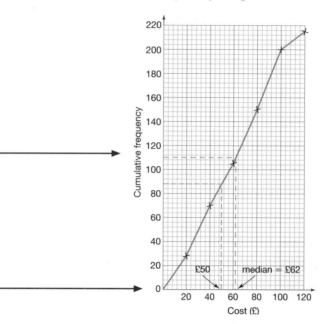

Plot each cumulative frequency value at the upper class boundary, i.e. 20, 40, 60…

Plot a value of zero at the lower boundary of the first class interval.

Draw a line from cumulative frequency $= \frac{215}{2} = 107.5$ to the curve.

d Use the graph to estimate the median price for the bottles of perfume sold that week.

Median cost = £62

e Use the graph to estimate the number of bottles of perfume sold that week that cost over £50.

Cumulative frequency for a cost of £50 is 88.

Number sold costing over £50 = 215 − 88 = 127

Look at 50 on the cost axis and read the cumulative frequency from the other axis.

DO**IT!**

Draw a flowchart for drawing a cumulative frequency graph.

CHECK**IT!**

1 The cumulative frequency table shows the waiting times in minutes for passengers to pass through security at an airport.

Time (t minutes)	Cumulative frequency
$0 < t \le 5$	15
$0 < t \le 10$	50
$0 < t \le 15$	122
$0 < t \le 20$	145
$0 < t \le 25$	165
$0 < t \le 30$	175
$0 < t \le 35$	180

a Plot a cumulative frequency graph for the information.

b Use your graph to find the median waiting time.

c Find

   i the upper quartile

   ii the lower quartile

   iii the interquartile range

d People who wait more than 27 minutes are often in danger of missing their flight. Work out what percentage of passengers waited 27 minutes or more at security. Give your answer to 1 decimal place.

Comparing sets of data

Box plots

Box plots are diagrams used to summarise important values in a set of data. They show:

- lower quartile
- median
- upper quartile
- lowest value
- highest value.

> Box plots give no information about individual values or how many values there are.

To measure the spread of a data set using a box plot, you can use:

- **range** = highest value − lowest value

- **interquartile range** (IQR) = upper quartile − lower quartile.

The interquartile range measures the spread of the middle half of the data. As it is the middle half of the data it will not be affected by extremely large or small values (**outliers**).

WORKIT!

The number of speeding fines issued in a city per week was recorded over a one-year period:

- lower quartile = 40
- median = 55
- upper quartile = 65
- lowest number = 30
- highest number = 80.

a Draw a box plot to summarise the data.

b Work out

  i the range

    Range = highest value − lowest value
           = 80 − 30 = 50

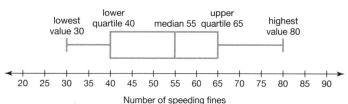

  ii the interquartile range.

    Interquartile range = upper quartile − lower quartile
                  = 65 − 40 = 25

Comparing data using box plots

Two sets of data can be compared using box plots. For example, these box plots show the results for a mock GCSE maths exam taken by two groups of students, A and B.

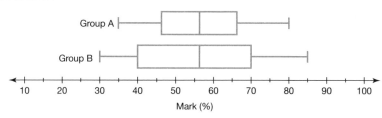

The box plots can be used to compare the two sets of data:

- Both groups have the same median mark of 56%.
- Group A has an upper quartile of 66% and a lower quartile of 46%, giving an interquartile range of 20%.
- Group B has an upper quartile of 70% and a lower quartile of 40% giving an interquartile range of 30%.
- The lower interquartile range of group A means that the marks were less spread and the middle half of the data was nearer to the median. This indicates that group A's results were more consistent.
- The range for group A is 80 − 35 = 45% and the range for group B is 85 − 30 = 55%. Therefore there was a greater variation in marks for group B.

Taking all these into account, there are reasons for arguing that group A's exam results were better than group B's.

Comparing data using cumulative frequency graphs and box plots

To compare sets of data you may be given two cumulative frequency graphs or a cumulative frequency graph for one set of data and a box plot for the other. From both, you can find the median, range, upper and lower quartiles and the interquartile range, and then compare the data sets.

DO IT!

Make a poster to show how two cumulative frequency graphs or box plots can be used to compare two sets of data.

CHECK IT!

1 The box plot and table give information about the distribution of heights of sunflowers in a garden.

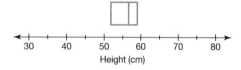

	Height (cm)
Lowest height	38
Lower quartile	
Median	
Upper quartile	
Highest height	69

Copy and complete the table and the box plot.

2 The cumulative frequency graphs show information about the marks 100 boys and 100 girls obtained in a maths exam.

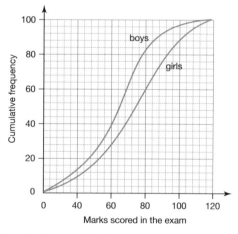

a For the boys, find
 i the range of the marks
 ii the median mark
 iii the upper quartile mark
 iv the lower quartile mark
 v the interquartile range of the marks.

b Using the median and the interquartile range, compare the distribution of the marks obtained by the boys with the marks obtained by the girls.

Scatter graphs

Types of correlation

Correlation is the relationship between variables on a scatter graph:

- **positive** correlation: the points slope upwards from bottom left to top right – as one quantity goes up, so does the other
- **negative** correlation: the points slope downwards from top left to bottom right – as one quantity goes up, the other goes down
- **no** correlation: the points do not slope in any definite direction – one quantity does not depend on the other.

Depending on how close the points are to making a straight line you can have:

- **strong** correlation, where the points are near to being in a straight line
- **weak** correlation, where the points are less closely grouped to a straight line.

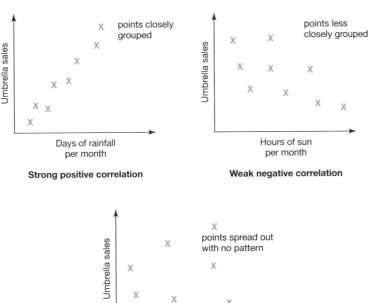

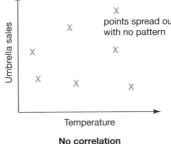

STRETCH IT!

Correlation between two quantities does not necessarily mean that one is caused by the other. For example, the increase in sales of umbrellas could be because there is a half-price sale and not because of wet weather.

Lines of best fit

The **line of best fit** is a line drawn between the points such that the total distance to the points either side of the line is approximately the same.

Sometimes a point is way out from the others. This is called an **outlier**. It represents an unusual value, or, in some cases, a mistake in the data. Ignore outliers when drawing your line of best fit.

When using a line of best fit to make estimates, extending the line past the last point could mean that the results are unreliable. You will be assuming that the trend shown by the line continues, which may or may not be the case.

WORKIT!

A supermarket sells wellington boots. Each day the shop assistant measures the amount of rainfall (cm) and the number of pairs of wellington boots sold, as shown on the scatter graph.

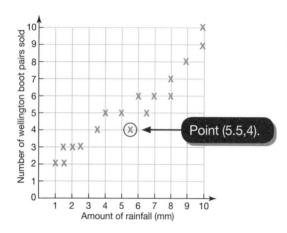

Point (5.5,4).

On a particular day there were 5.5mm of rainfall and 4 pairs of wellington boots were sold.

a On the scatter graph, circle the point representing the data for this day.

b Describe the relationship between the amount of rainfall and the number of pairs of wellington boots sold.

Strong positive correlation: as the rainfall goes up so does the number of wellington boots sold.

c One day had 3mm of rainfall. Estimate the number of pairs of wellington boots sold.

3 ← Draw in the line of best fit. Read the information off your line, at the 3 mm value on the horizontal axis.

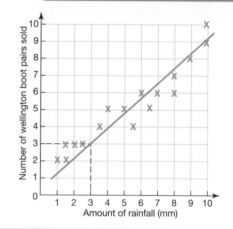

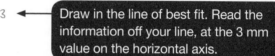

DOIT!

Draw a poster showing different types of correlation.

CHECKIT!

1 The table shows the daily number of visitors to a beach and the temperature at noon.

Temperature (°C)	5	10	15	20	25	30	35
Daily visitors	17	30	40	53	60	85	90

a Draw a scatter graph for the data, with temperature on the horizontal axis.

b Draw the line of best fit.

c Use your line of best fit to estimate the number of visitors to the beach when the temperature is 27 °C.

d i Use your line of best fit to estimate the number of visitors to the beach when the temperature is 45 °C.

  ii Explain why this estimate of the number of visitors is likely to be unreliable.

1 The table shows information about the amount of pocket money per week each student in a class receives.

Amount of pocket money (£a per week)	Number of students
$2 < a \leq 4$	6
$4 < a \leq 6$	15
$6 < a \leq 8$	9
$8 < a \leq 10$	6

a How many students are there in the class?

b Work out an estimate for the mean amount of pocket money for the class.

c Write down the modal class interval.

2 The table shows the marks 100 students got in an exam.

Mark (m)	Frequency
$0 < m \leq 20$	6
$20 < m \leq 40$	19
$40 < m \leq 60$	39
$60 < m \leq 80$	21
$80 < m \leq 100$	15

a Draw the cumulative frequency curve for this information.

b Use your graph to work out the median mark for the class.

It has been decided to set the pass mark so that only 75% of the students pass the exam.

c Use the graph to find the minimum mark needed to pass the exam.

3 Four schools, A, B, C and D, each have 150 students. They take the same examination, with a maximum mark of 80. The cumulative frequency graphs for each school are shown.

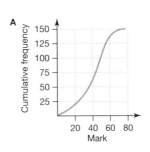

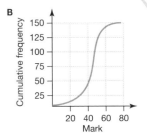

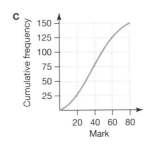

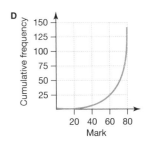

a Which school had the highest median mark?

b Which school had the largest interquartile range?

c Which school performed best in the exam? Give reasons for your answer.

4 Alex and Nadia are organising a street party for 320 people. They interview a sample of 40 people in the street.

a How should they select their sample?

b One question they asked was about what people would drink. The table shows their results.

Drink	Number of people
Cola	7
Lemonade	15
Squash	5
Tea	10
Coffee	3

How many people should they buy lemonade for?

Answers

Number

Integers, decimals and symbols

1 **a** 200.1   **b** 2.001   **c** 2.3   **d** 87
2 **a** 140.94   **b** 1.4094   **c** 290   **d** 4.86
3 −0.5, 0, 0.012, 0.12, 12
4 **a** $\frac{5}{0.5} = 10$ **b** $1\frac{5}{9} > \frac{4}{3}$ **c** $-3 < -1$
5 **a** 5   **c** −15   **e** −4
   b −8   **d** 6

Addition, subtraction, multiplication and division

1 **a** 1561   **c** 69.93
   b 3047   **d** 23.923
2 **a** 2819   **c** 8.185
   b 287   **d** 5.401
3 **a** 29798   **b** 29.26   **c** 40.768
4 **a** 46   **b** 343   **c** 35.4

Using fractions

1 **a** $\frac{16}{5} = 3\frac{1}{5}$ **c** $\frac{5}{8}, \frac{3}{4}, \frac{9}{10}, 1\frac{1}{5}, \frac{16}{5}$
   b $1\frac{1}{5} = \frac{6}{5}$ **d** $4\frac{2}{5}$ **e** $2\frac{23}{40}$
2 $\frac{15}{45}, \frac{4}{12}, \frac{16}{48}$
3 **a** $7\frac{1}{3}$ **b** $3\frac{1}{2}$
4 $\frac{13}{60}$

Different types of number

1 **a** 16   **b** 5   **c** 16
2 $2 \times 2 \times 3 \times 5 \times 5$
3 every 144 days
4 **a** $2^2 \times 3^3 \times 7$ **b** 36

Listing strategies

1 12   2 180

The order of operations in calculations

1 **a** 18   **b** 13   **c** 25
2 **a** 10   **b** 23   **c** 5

Indices

1 **a** 7^{10} **b** 3^{-6} **c** 5^{20}
2 **a** 5^7 **c** $2^{10} \times 5^{-3}$
   b 6^{-3} **d** $7^{10} \times 11^{-1}$
3 **a** 1   **c** 16   **e** 18
   b 10   **d** $\frac{1}{5}$ (or 0.2)
4 $x = 1$

Surds

1 **a** $\sqrt{6}$ **b** 5   **c** 18   **d** 20
2 $a = 2$
3 **a** $3\sqrt{5}$ **b** $6\sqrt{2}$
4 **a** $\frac{16}{3\sqrt{2}} = \frac{16\sqrt{2}}{3\sqrt{2}\sqrt{2}} = \frac{16\sqrt{2}}{3 \times 2} = \frac{8\sqrt{2}}{3}$ **b** $8 - 2\sqrt{7}$
5 **a** −4   **b** $7 + 4\sqrt{3}$ **c** $5 + 3\sqrt{3}$

Standard form

1 **a** 0.005   **b** 565000
2 **a** 2.5×10^4 **c** 5×10^2
   b 1.25×10^{-3} **d** 1.4×10^{-2}
3 **a** 9×10^{-4} **c** 2×10^2
   b 2.4×10^3 **d** 8.04×10^4
4 **a** 1.33×10^{10} pounds
   b 26600000 people
5 **a** 1.55×10^4 **c** 5×10^2
   b 655000   **d** 4×10^3

Converting between fractions and decimals

1 **a** 0.43   **b** 0.375   **c** 0.55
2 **a** $\frac{4}{5}$ **b** $\frac{9}{20}$ **c** $\frac{73}{125}$
3 **a** $\frac{7}{9}$ **b** $\frac{2}{45}$ **c** $\frac{21}{22}$
4 **a** $\frac{14}{27}$ **b** $\frac{19}{25}$
5 1

Converting between fractions and percentages

1 **a** $\frac{1}{4}$ **b** $\frac{17}{20}$ **c** $\frac{17}{25}$
2 maths: 81.25%
   Charlie did better at maths.
3 **a** 30%   **b** 16%   **c** 42.9%

Fractions and percentages as operators

1 **a** £480   **b** £4.50   **c** 76kg
2 School A: 336, School B: 455

Standard measurement units

1 **a** 9700g   **b** 0.85 litres   **c** 205000cm
2 8.64×10^4 seconds
3 £81.60

Rounding numbers

1 **a** 1260   **c** 0.000308   **e** 1.81×10^{-4}
   b 14.9   **d** 9080000
2 **a** 10.6   **c** 0.03   **e** 0.002
   b 123.977   **d** 3.971   **f** 4.10
3 **a** 2000   **b** 2000   **c** 1990
4 0.0004 (4 decimal places)

Estimation

1 a 2400   **b** 3

2 a A   **e** A   **i** B
   b C   **f** B   **j** A
   c B   **g** C
   d B   **h** A

3 a 6.7 – accept 6.5 to 6.9
   b 10.2 – accept 10.1 to 10.3
   c 12 – accept 10 to 14
   d 5 – accept 4 to 6

Upper and lower bounds

1 a 144.5 cm
   b 145.5 cm
   c $144.5 \leq l < 145.5$ cm

Review it!

1 a 24 647.515   **b** 21.5 (to 1 d.p.)

2 a 6.5   **c** 81
   b 4   **d** 5.94

3 a $5\frac{1}{3}$ **b** $9\frac{7}{13}$

4 $4\frac{3}{4} + \frac{1}{2}$

5 a lower bound = 10.113 cm² (to 3 d.p.)
     upper bound = 10.180 cm² (to 3 d.p.)
   b 10 cm²

6 a 3.34×10^{23} molecules
   b 2.99×10^{-26} kg

7 0.16

8 a 1   **c** $\frac{1}{64}$
   b 3   **d** $\frac{1}{4}$

9 $5.55 \leq y < 5.65$

10 $\frac{1-\sqrt{2}}{1+\sqrt{2}} = \frac{1-\sqrt{2}}{1+\sqrt{2}} \times \frac{1-\sqrt{2}}{1-\sqrt{2}} = \frac{1-2\sqrt{2}+2}{1-2} = \frac{3-2\sqrt{2}}{-1} = 2\sqrt{2} - 3$

11 $111.5 \leq a < 112.5$

12 $\frac{8}{11}$

13 a $0.287996 \leq c < 0.289272$
   b 0.29 (to 2 s.f.)

14 a $\frac{2}{3}$
   b 80

15 a 4.5×10^{-7}
   b 1.2×10^{7}
   c 5.64×10^{3}

16 3.2×10^{-1}

17 a 1, 2, 4, 8, 16, 32, 64
   b 4

Algebra

Simple algebraic techniques

1 a formula   **b** identity   **c** expression
   d identity   **e** equation

2 a $10x^2 + 4x$ **c** $-3x^2 + 10xy$
   b $3a - b$ **d** $3x^3 - x - 5$

3 16

4 4

Removing brackets

1 a $2x + 8$ **c** $x - 1$ **e** $3x^2 + 3x$
   b $63x + 21$ **d** $3x^2 - x$ **f** $20x^2 - 8x$

2 a $5x + 12$ **c** $4x^2 + 2x$
   b $3x + 45$ **d** $3x^2 - 10x + 8$

3 a $t^2 + 8t + 15$ **c** $6y^2 + 41y + 63$
   b $x^2 - 9$ **d** $4x^2 - 4x + 1$

4 a $2x^3 + 21x^2 + 55x + 42$
   b $24x^3 - 46x^2 + 29x - 6$

Factorising

1 a $6(4t + 3)$ **c** $5y(x + 3z)$
   b $a(9 - 2b)$ **d** $6xy^2(4x^2 + 1)$

2 a $(x + 7)(x + 3)$ **c** $(2x + 5)(3x + 2)$
   b $(x + 5)(x - 3)$ **d** $(2x + 7)(2x - 7)$

3 $\frac{1}{2x + 3}$

Changing the subject of a formula

1 a $r = \sqrt{\frac{A}{\pi}}$ **b** $r = \sqrt{\frac{A}{4\pi}}$ **c** $r = \sqrt[3]{\frac{3V}{4\pi}}$

2 a $c = y - mx$ **d** $s = \frac{v^2}{2a}$
   b $u = v - at$ **e** $u = \sqrt{v^2 - 2as}$
   c $a = \frac{v-u}{t}$ **f** $t = \frac{2s}{u + v}$

Solving linear equations

1 a $x = 3$ **b** $x = 3$ **c** $x = 20$

2 a $x = 5$ **b** $x = 18$ **c** $x = 20$

3 a $x = -2$ **b** $m = 1$ **c** $x = \frac{6}{5}$, $1\frac{1}{5}$ or 1.2

Solving quadratic equations using factorisation

1 a $x = -2$ or $x = -3$
   b $x = -3$ or $x = 4$
   c $x = -\frac{7}{2}$ or $x = -5$

2 a Area = $\frac{1}{2} \times$ base $\times$ height
      $\frac{1}{2}(2x + 3)(x + 4) = 9$
      $\frac{1}{2}(2x^2 + 11x + 12) = 9$
       $2x^2 + 11x + 12 = 18$
       $2x^2 + 11x - 6 = 0$
   b $x = \frac{1}{2}$ **c** base = 4 cm, height = 4.5 cm

3 By Pythagoras' theorem $(x + 1)^2 + (x + 8)^2 = 13^2$

$x^2 + 2x + 1 + x^2 + 16x + 64 = 169$

$2x^2 + 18x - 104 = 0$

Dividing through by 2 gives

$x^2 + 9x - 52 = 0$

$(x - 4)(x + 13) = 0$

So $x = 4$ or -13 (disregard $x = -13$ as x is a length)

Hence $x = 4$ cm

Solving quadratic equations using the formula

1 $x = 2.14$ or $x = -1.64$ (to 3 s.f.)

2 **a** $\frac{2x + 3}{x + 2} = 3x + 1$

$\qquad 2x + 3 = (3x + 1)(x + 2)$

$\qquad 2x + 3 = 3x^2 + 7x + 2$

$\qquad 0 = 3x^2 + 5x - 1$

$\quad 3x^2 + 5x - 1 = 0$

$\quad$ **b** $x = -1.85$ or $x = 0.18$ (to 2 d.p.)

Solving simultaneous equations

1 **a** $x = 2, y = -1$

$\quad$ **b** $x = 4, y = 2$

2 $x = \frac{1}{5}, y = -2\frac{3}{5}$ or $x = \frac{1}{2}, y = -2$

3 Equating the y values gives

$x^2 + 5x - 4 = 6x + 2$

$x^2 - x - 6 = 0$

$(x - 3)(x + 2) = 0$

$x = 3$ or -2

When $x = 3, y = 6 \times 3 + 2 = 20$

When $x = -2, y = 6 \times (-2) + 2 = -10$

Points are $(3, 20)$ and $(-2, -10)$

Solving inequalities

1 **a** $x > 6$

$\quad \{x: x > 6\}$

$\quad$ **b** $x \geq 11$

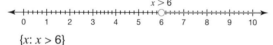

$\quad \{x: x \geq 11\}$

$\quad$ **c** $x < 26$

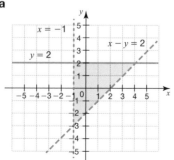

$\quad \{x: x < 26\}$

2 **a** $x > 10$ $\quad$ **b** $x < 0.4$ or $\frac{2}{5}$ $\qquad$ **c** $x \leq 8$

3 **a**

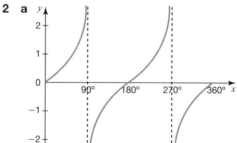

$\quad$ **b** $(0, 2), (0,1), (0, 0), (0, -1), (1, 2), (1, 1), (1, 0), (2, 2),$
$\qquad (2, 1), (3, 2)$

4 $x < -2$ and $x > 5$

Problem solving using algebra

1 26, 51 $\qquad\qquad$ **2** 65 $\qquad\qquad$ **3** 9 cm by 3 cm

Use of functions

1 **a** -1 $\qquad$ **b** $-\frac{2}{3}$ $\qquad$ **c** $\frac{x + 1}{x}$

2 **a** $\sqrt{x^2 + 8x + 7}$ $\qquad\qquad\qquad$ **c** 4

$\quad$ **b** $\sqrt{(x^2 - 9)} + 4$

Iterative methods

1 1.521

$\quad$ Let $f(x) = x^3 - x - 2$

$\qquad f(1.5215) = (1.5215)^3 - 1.5215 - 2 = 0.0007151$

$\qquad f(1.5205) = (1.5205)^3 - 1.5205 - 2 = -0.005225$

$\quad$ As there is a change in sign, $a = 1.521$ to 3 decimal places.

Equation of a straight line

1 **a** 2 $\qquad$ **b** $-\frac{1}{2}$ $\qquad$ **c** $y = -\frac{1}{2}x + 5$

2 $y = 3x - 3$

3 $2x - y + 2 = 0$

4 **a** $\frac{1}{2}$ $\qquad$ **b** $(2, 2)$ $\qquad$ **c** **i** -2

$\qquad\qquad\qquad\qquad\qquad\qquad\qquad\quad$ **ii** $y = -2x + 6$

Quadratic graphs

1 **a** $2(x - 3)^2 - 17$ $\qquad$ **c**

$\quad$ **b** **i** $(3, -17)$

$\qquad\quad$ **ii** $x = 0.1$ and $x = 5.9$

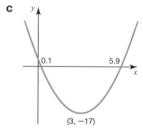

2 **a** $y = x^2 - 4x - 5$ $\qquad$ **b** $y = -x^2 + 9x - 14$

3 **a** $x^2 + 12x - 16 = (x + 6)^2 - 52$

$\quad$ **b** $(-6, -52)$

Recognising and sketching graphs of functions

1 **a** B $\qquad$ **c** E $\qquad$ **e** D

$\quad$ **b** F $\qquad$ **d** A $\qquad$ **f** C

2 **a**

$\quad$ **b** $x = 240°$

3 **a** A $\qquad$ **b** G $\qquad$ **c** F $\qquad$ **d** E

Translations and reflections of functions

1 a (3, 5) **c** (2, −5)
 b (−1, 5) **d** (−2, 5)

2 a

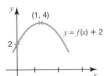

c

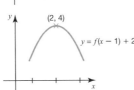

b

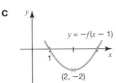

d

Equation of a circle and tangent to a circle

1 a (0, 0) **b** 7

2 a $x^2 + y^2 = 100$

 b Gradient of radius to (8, 6) $= \frac{6}{8} = \frac{3}{4}$
  Gradient of tangent $= -\frac{4}{3}$

 c $y = -\frac{4}{3}x + 16\frac{2}{3}$

Real-life graphs

1 a 5 km/h **b** 0.25 hours **c** 24 km/h

2 a

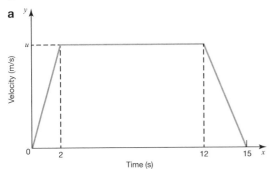

 b $u = 4$ m/s **c** 1.33 m/s²

Generating sequences

1 a 17 **c** −12 **e** $\frac{1}{48}$
 b 3.0 **d** 432 **f** $-\frac{1}{16}$

2 17, 290

3 2.25, 5.5

The *n*th term

1 a 47, 44, 41
 b no **c** −1

2 a 6, 18, 54, 162
 b Both 2 and 3 are factors, so 6 must also be a factor.

3 a *n*th term $= 2n - 3$ **b** $x = 31$

4 *n*th term $= 4n^2 + n - 1$

Arguments and proofs

1 a True: $2n$ is always even as it is a factor of 2. Adding 1 to an even number always gives an odd number.
 b False: $x^2 = 9$, so $x = \sqrt{9} = \pm 3$.

c False: n could be a decimal such as 4.25, so squaring it would not give an integer.

d False: if $n \leq 1$ this is not true.

2 Let the four consecutive numbers be n, $n + 1$, $n + 2$ and $n + 3$.

 Sum $= n + (n + 1) + (n + 2) + (n + 3)$
  $= 4n + 6$
  $= 2(2n + 3)$

 Since the sum is a multiple of 2, it is always even.

 Therefore the sum of four consecutive numbers is always even.

3 Let the consecutive integers be x, $x + 1$ and $x + 2$.

 Sum of the integers $= x + x + 1 + x + 2 = 3x + 3$
 $= 3(x + 1)$

 As 3 is a factor, the sum must be a multiple of 3.

4 a The numerator is larger than the denominator so the fraction will always be greater than 1. Statement is false.

 b As a is larger than b, squaring a will result in a larger number than squaring b. Hence $a^2 > b^2$ so the statement is false.

 c The square root of a number can have two values one positive and the other negative so this statement is false.

Review it!

1 a $-9x + 12$ **b** $6x + 4$ **c** $6x^3 + 25x^2 + 16x - 15$

2 a $(2x - 1)(x + 4)$ **b** $x = \frac{1}{2}$ or $x = -4$

3 a $8x^6y^3$ **b** $6x$ **c** $\frac{5}{b}$

4 $x = 2$, $y = 1$

5 a $\frac{3}{x + 7} = \frac{2 - x}{x + 1}$
  $3(x + 1) = (2 - x)(x + 7)$
  $3x + 3 = 2x + 14 - x^2 - 7x$
  $3x + 3 = -x^2 - 5x + 14$
  $x^2 + 8x - 11 = 0$
 b $x = 1.20$ or $x = -9.20$ (to 2 d.p.)

6 $x = \frac{3y - 2z}{az + 1}$

7 a $f^{-1}(x) = 3x - 15$ or $f^{-1}(x) = 3(x - 5)$
 b $k = 7$

8 a 26, 22, 18 **b** 8th term $= -2$

9 The point lies outside the circle.

10 $x - 9y$

11 a $2(x + 2)^2 - 7$
 b **i** (−2, −7)
  ii $x = -3.9$ and $x = -0.1$
 c

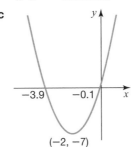

12 30 cm²

13 $y = 7x + 12$

14 $x = -2, y = 0$ or $x = \frac{6}{5}, y = \frac{8}{5}$

Ratio, proportion and rates of change

Introduction to ratios

1 **a** 1:3 **b** 5:12 **c** 4:9

2 **a** 1:8 **b** 100:1 **c** 1:50

3 £250, £150

4 315 members

5 £42 000

6 62.9%

7 £5.70

8 10:5:1

Scale diagrams and maps

1 30 cm

2 **a** 1:500 000 **b** 6 km

Percentage problems

1 2.67% **3** £39 330 **5** £2520

2 88.3% **4** £356

Direct and inverse proportion

1 Inverse proportion means that if one quantity doubles the other quantity halves.

2 **a** $y = kx$ **b** 10.7

3 **a** cheaper in the UK

   b £5.49 cheaper in the UK

4 268 cm³

5 33 333 (to nearest whole number)

6 150

7 **a** $A = 6x^2, k = 6$ **b** 96 cm²

Graphs of direct and inverse proportion and rates of change

1 Graph C

2 Graph B

3 $a = 6$

4 Equation connecting x and y is $y = \frac{k}{x}$

   When $x = 1, y = 4$ so $4 = \frac{k}{1}$ so $k = 4$

   The equation of the curve is now $y = \frac{4}{x}$

   When $x = 4, y = \frac{4}{4} = 1$ so $a = 1$

   When $y = 0.8, 0.8 = \frac{4}{x}$ giving $x = 5$ so $b = 5$.

   Hence $a = 1$ and $b = 5$.

5 $y = kx^2$

   When $x = 2, y = 16$ so $16 = k \times 2^2$ giving $k = 4$.

   $y = 4x^2$

   Hence $36 = 4x^2$ so $x = 3$

   $a = 3$

Growth and decay

1 **a** 1.05 **c** 1.0375

   b 1.25 **d** 0.79

2 £4962 (to nearest whole number)

3 2048

Ratios of lengths, areas and volumes

1 $6\frac{2}{3}$ cm

2 $\frac{V_A}{V_B} = \frac{27}{64} =$ (scale factor)³, so scale factor $= \frac{3}{4}$

   $\frac{A_A}{A_B} =$ (scale factor)² $= \frac{9}{16}$

   So $A_A = \frac{9}{16} \times 96 = 54$ cm²

3 **a** Triangles ABE and ACD must be proved similar:

   BE parallel to CD, all the corresponding angles in both triangles are the same.

   $BE = 4$ cm

   b 10 cm²

Gradient of a curve and rate of change

1 **a** The gradient represents the acceleration.

   b 5 m/s²

   c 3.9 s

Converting units of areas and volumes, and compound units

1 **a** **i** 14 800 mm²

   ii 0.0148 m²

   b 0.000 12 m³

2 1.932 g/cm³ (to 3 d.p.)

3 2 000 000

4 13.8 m/s (to 2 d.p.)

5 **a** 56 km/h

   b The distance will not be the same, so the average speed will be different.

Review it!

1 £96

2 **a** 5.10 m²

   b 5.31 m²

3 $y = 2$

4 243 students

5 £485 000

6 **a** £2250

   b £2262.82

7 **a** 3420 yuan

   b travel agent

8 £30 000

9 0.64 cm³

10 20%

11 8.7 g/cm³

Geometry and measures

2D shapes

1 a true   **c** true   **e** true
   b false   **d** false   **f** true

Constructions and loci

1

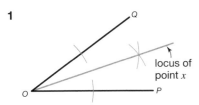

locus of point x

2 This is a reduced version of the answer.

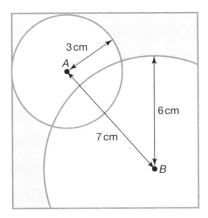

Properties of angles

1 144°

2 a 9   **b** 1260°

3 20°

4 a x = angle EBC = 55° (alternate angles)

   b Angle EHI = 180 − (85 + 55) = 40° (angle sum in a triangle)

      Angle DEH = angle EHI = 40° (alternate angles)

5 77°, 103°, 77°, 103° (a = 24).

Congruent triangles

1 $AC = AC$ (common to both triangles)
   $AB = AD$ (given)
   $BC = CD$ (given)
   Triangles ACD and ACB are congruent (SSS).
   Hence angle ABC = angle ADC

2 $AB = AC$ (given)
   $BM = MC$ (M is the midpoint of BC)
   $AM = AM$ (common to both triangles)
   Triangles ABM and ACM are congruent (SSS).
   Hence angle AMB = angle AMC
   Angle AMB + angle AMC = 180°, so angle
   $AMB = \frac{180}{2}$ = 90°

Transformations

1 a, b

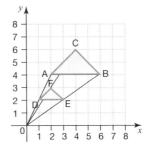

   c Enlargement, scale factor 2, centre of enlargement (0, 0)

2

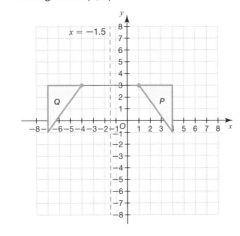

3

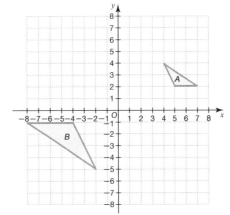

4 Translation by $\binom{5}{-4}$

Invariance and combined transformations

1 a, b, c

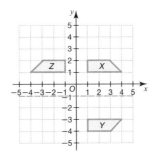

   d Reflection in the y-axis

2 a, b

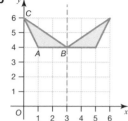

c (3, 4)

3 a, b

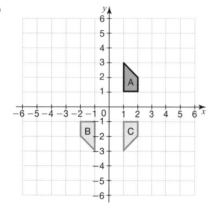

c A reflection in the *x*-axis.

3D shapes

1

Shape	Number of vertices	Number of faces	Number of edges
Triangular-based pyramid	4	4	6
Cone	1	2	1
Cuboid	8	6	12
Hexagonal prism	12	8	18

2 $V = 16$, $F = 10$, $E = 24$; $V + F - E = 16 + 10 - 24 = 2$

Parts of a circle

1 a radius   **c** minor arc   **e** minor sector
   b chord   **d** minor segment

Circle theorems

1 a angle ACB = angle YAB = 30° (alternate segment theorem)

   b angle ABC = 90° (angle in a semicircle)

   c angle ADC = 90° (angle in a semicircle)

2 a angle OAX = 90° (angle between tangent and radius)

   b angle AOX = 180 − (90 + 30) = 60° (angle sum in a triangle)

   c angle ACB = 30° (angle at centre twice angle at circumference)

3 a angle ADB = 40° (angles on same arc)

   b angle EDB = angle EDA + angle ADB = 50 + 40 = 90°

    So BD is a diameter of the circle (angle between tangent and diameter is 90°)

    angle BAD = 90° (angle in a semicircle)

4 Angle ACB = 46° (angles bounded by the same chord in the same segment are equal)

Angle ABC = 90° (angle in a semi-circle is a right-angle)

Angle BAC = 180 − (90 + 46) = 44° (angles in a triangle add up to 180°)

Projections

1 a

   c

   b

2

Bearings

1 245°

2 a 225°   **b** 320°

Pythagoras' theorem

1 a 10.3 cm (to 1 d.p.)   **b** 8.8 cm (to 1 d.p.)

2 height = 10.91 cm

area = 54.54 cm²

3 x = 8.12 cm (to 2 d.p.)

4 Let the perpendicular height of the triangle = h

Area of triangle = $\frac{1}{2} \times 14 \times h$

Hence $\frac{1}{2} \times 14 \times h$ = 90

Solving gives h = 12.8571 cm

By Pythagoras' theorem, $AC^2 = 7^2 + 12.8571^2$

Solving gives AC = 14.6 cm (3 significant figures)

Area of 2D shapes

1 £1596

2 $\frac{5}{8}$

3 a area of semicircle = $\frac{1}{2} \times \pi \times x^2 = \frac{\pi x^2}{2}$

    area of rectangle = $4x \times 2x = 8x^2$

    area of shape = $\frac{\pi x^2}{2} + 8x^2 = x^2(8 + \frac{\pi}{2})$

   b $x(10 + \pi)$

Volume and surface area of 3D shapes

1 $r = \sqrt{\frac{3}{2}a^3}$

2 a 12.5 m²   **b** 62.5 m³   **c** 10 hours (to nearest hour)

3 1200 m² (to 3 s.f.)

Trigonometric ratios

1 a 17.32 cm (to 2 d.p.)

b 9.19 cm (to 2 d.p.)

2 a 73.3° (to 0.1°)

b 50.3° (to 0.1°)

3 $\sin\theta = \frac{b}{c}$ and $\cos\theta = \frac{a}{c}$

$\frac{\sin\theta}{\cos\theta} = \frac{\frac{b}{c}}{\frac{a}{c}} = \frac{b}{c} \times \frac{c}{a} = \frac{b}{a} = \tan\theta$

4 a 13.60 cm (to 2 d.p.)

b 63.8° (to 1 d.p.)

c 8.56 cm (to 2 d.p.)

Exact values of sin, cos and tan

1 $\sqrt{3}\tan 30° + \cos 60° = \sqrt{3} \times \frac{1}{\sqrt{3}} + \frac{1}{2} = 1 + \frac{1}{2} = \frac{3}{2}$

$a = 3, b = 2$

2 $\sin 30 = \frac{1}{2}$ and $\cos 30 = \frac{\sqrt{3}}{2}$

$\sin^2 30 = \frac{1}{4}$ and $\cos^2 30 = \frac{3}{4}$

$\sin^2 30 + \cos^2 30 = \frac{1}{4} + \frac{3}{4} = 1$

Sectors of circles

1 47.7° (to 1 d.p.)

2 a 7.99 cm   **b** 3.94 cm²

3 Radius $OA = 10.3$ cm

4 Area of sector = 109 cm²

Sine and cosine rules

1 a 15 cm²

b 15.5 cm (to 3 s.f.)

2 27.7° (to 0.1°)

3 a 38.6°

b If angle *XZY* is not obtuse, then angle *XYZ* can be obtuse. An alternative answer would be 141.4°.

Vectors

1 a $\begin{pmatrix} -6 \\ 8 \end{pmatrix}$ **b** $\begin{pmatrix} 2 \\ 0 \end{pmatrix}$

2 a $\mathbf{b} - \mathbf{a}$ **b** $\frac{3}{5}(\mathbf{b} - \mathbf{a})$ **c** $\frac{1}{5}(2\mathbf{a} + 3\mathbf{b})$

3 a $-\mathbf{a} - 3\mathbf{b}$

b $\overrightarrow{BM} = \frac{1}{2}\overrightarrow{BC} = \frac{1}{2}(-\mathbf{a} - 3\mathbf{b})$

$\overrightarrow{PM} = \overrightarrow{PA} + \overrightarrow{AB} + \overrightarrow{BM} = 2\mathbf{b} + \mathbf{a} + \frac{1}{2}(-\mathbf{a} - 3\mathbf{b})$

$= \frac{1}{2}\mathbf{b} + \frac{1}{2}\mathbf{a} = \frac{1}{2}(\mathbf{a} + \mathbf{b})$

$\overrightarrow{MD} = \overrightarrow{MB} + \overrightarrow{BD} = -\frac{1}{2}(-\mathbf{a} - 3\mathbf{b}) + \mathbf{a} = \frac{3}{2}\mathbf{a} + \frac{3}{2}\mathbf{b}$

$= \frac{3}{2}(\mathbf{a} + \mathbf{b})$

$\overrightarrow{PM}$ and $\overrightarrow{MD}$ have the same vector part $(\mathbf{a} + \mathbf{b})$, therefore they are parallel. Both lines pass through M and parallel lines cannot pass through the same point unless they are the same line. Hence *PMD* is a straight line.

Review it!

1 Angle *ADC* = (180 − *x*)° (opposite angles of cyclic quadrilateral)

Angle *ADE* = *y*° (alternate segment theorem)

Angle *CDF* = 180 − angle *ADE* − angle *ADC*
= 180 − *y* − (180 − *x*) = (*x* − *y*)°

2

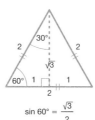

$\sin 60° = \frac{\sqrt{3}}{2}$ $\cos 45° = \frac{1}{\sqrt{2}}$

$\cos 45° + \sin 60° = \frac{1}{\sqrt{2}} + \frac{\sqrt{3}}{2}$

$= \frac{1}{\sqrt{2}} \times \frac{\sqrt{2}}{\sqrt{2}} + \frac{\sqrt{3}}{2}$

$= \frac{\sqrt{2}}{2} + \frac{\sqrt{3}}{2}$

$= \frac{1}{2}(\sqrt{2} + \sqrt{3})$

3 5, 12 and 13

4 a

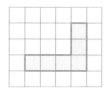

b

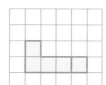

5 a

b i

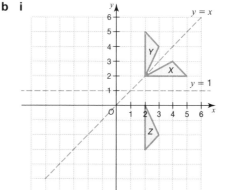

b ii (2, 2)

c See triangle *Z* on grid above

d Rotation of 90° clockwise about (1, 1)

6 a 19.4 km (to 1 s.f.)

b 280°

Probability

The basics of probability

1 $\frac{8}{125}$

2 a 0

  b $\frac{1}{5}$

  c $\frac{11}{20}$

Probability experiments

1 a 0.21 (to 2 d.p.)

  b 0.14 (to 2 d.p.)

  c Sean is wrong. 120 spins is a small number of spins and it is only over a very large number of spins that the relative frequencies may start to be nearly the same.

2 a 0.04   **b** 600 cans

3 45 apples

The AND and OR rules

1 a Independent events are events where the probability of one event does not influence the probability of another event occurring. Here it means that the probability of the first set of traffic lights being red does not affect the probability of the second set being red.

  b 0.06   **c** 0.56

2 a $\frac{7}{48}$ **b** $\frac{1}{32}$

Tree diagrams

1 a $\frac{1}{15}$

  b $= \frac{7}{15}$

2 20 balls

3 a

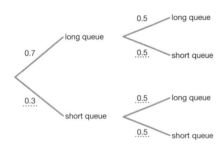

  b 0.15

  c 0.85

4 a

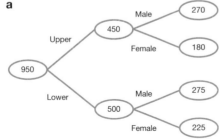

  b $\frac{109}{190}$ or 0.57

Venn diagrams and probability

1 a i {1, 3, 4, 5 8, 9, 10, 11}

    ii {8, 9}

    iii {2, 5, 6. 10, 13}

  b $\frac{7}{11}$

2

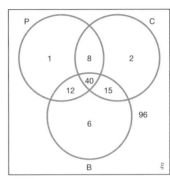

  a $\frac{10}{21}$

  b $\frac{3}{28}$

  c $\frac{48}{61}$

3 a

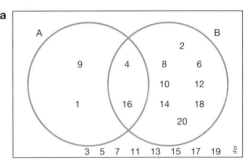

  b $P = \frac{8}{20} = \frac{2}{5}$

Review it!

1 a

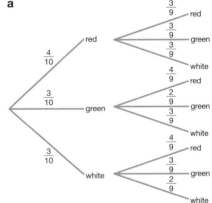

  b $\frac{4}{15}$

  c $\frac{11}{15}$

2 a

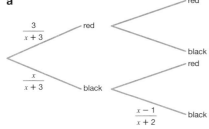

P(two black) $= \left(\frac{x}{x+3}\right) \times \left(\frac{x-1}{x+2}\right) = \frac{7}{15}$

$15x(x - 1) = 7(x + 3)(x + 2)$

$15x^2 - 15x = 7x^2 + 35x + 42$

$8x^2 - 50x - 42 = 0$

$4x^2 - 25x - 21 = 0$

b 10 balls

c $\frac{7}{15}$

3

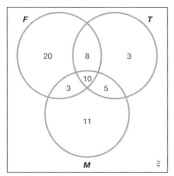

a $\frac{11}{60}$ **b** $\frac{15}{29}$

4 a

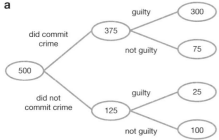

b $\frac{13}{20}$ **c** $\frac{1}{5}$

Statistics

Sampling

1 26 **2** 11

Two-way tables, pie charts and stem-and-leaf diagrams

1 a

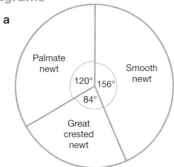

b No. The other pond might have had more newts in total. The proportion of smooth newts in the second pond is lower, but there may be more newts.

2

2	2	3	3	4		
3	0	0	1	3	5	5	6
4	0	1	2	5	7		

key 2|2 represents 22 mins

Line graphs for time series data

1 a 926, 904, 885, 854

b The trend is decreasing sales.

2 a

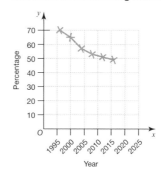

b The percentage of people using the local shop is decreasing.

c 44%

d There are no points so you would be trying to predict the future. There may be a change of ownership/or a refurbishment making it more popular. It could even close down before then.

Averages and spread

1 13.5 years

2 Mean $= \frac{\text{total number of marks}}{\text{number of students}}$

Total mark for boys $= 50 \times 10 = 500$

Total mark for girls $= 62 \times 15 = 930$

Total mark for class $= 500 + 930 = 1430$

Mean for class $= \frac{\text{total number of marks}}{\text{number of students}} = \frac{1430}{25} = 57.2\%$

Joshua is wrong, because he didn't take account of the fact that there was a different number of boys and girls.

3 a

Cost (£C)	Frequency	Mid-interval value	Frequency × mid-interval value
$0 < C \le 4$	12	2	24
$4 < C \le 8$	8	6	48
$8 < C \le 12$	10	10	100
$12 < C \le 16$	5	14	70
$16 < C \le 20$	2	18	36

b £7.51 (to nearest penny)

Histograms

1

Mass (m pounds)	Frequency
$0.0 < m \le 0.5$	30
$0.5 < m \le 1.0$	16
$1.0 < m \le 1.5$	13
$1.5 < m \le 2.0$	21
$2.0 < m \le 2.5$	32
$2.5 < m \le 3.0$	38

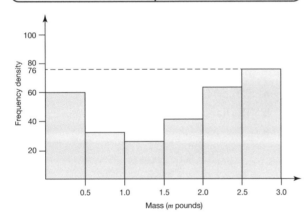

Cumulative frequency graphs

1 a

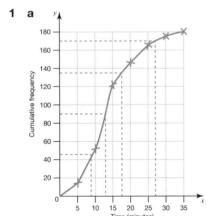

b 13 minutes

c   **i** 17.5 minutes

    ii 9 minutes

    iii 8.5 minutes

d 5.6% (to 1 d.p.)

Comparing sets of data

1

	Height (cm)
Lowest height	38
Lower quartile	**52**
Median	**57**
Upper quartile	**59**
Highest height	69

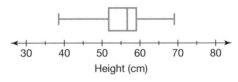

Height (cm)

2 a   **i** 120 marks

    ii 65 marks

    iii 75 marks

    iv 51 marks

    v 24 marks

b For the girls: median mark = 74 marks, upper quartile = 89 marks, lower quartile = 58 marks and interquartile range = 31 marks

Comparison:

The median mark for the girls is higher (or higher average mark).

The interquartile range is lower for the boys showing that their marks are less spread out for the middle half of the marks.

Scatter graphs

1 a, b

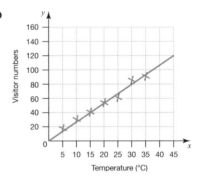

c 71 visitors

d **i** 120 visitors

   ii There are no points near this temperature so you cannot assume the trend continues.

Review it!

1 a 36 students   **c** $4 < a \leq 6$

   b £5.83 to the nearest penny

2 a

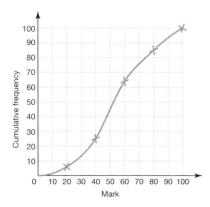

   b 54   **c** 40

3 a D (the median mark is furthest to the right.)

   b C (the largest gap between the quartiles.)

   c D as the median mark is the highest and also the interquartile range is small which means 50% of pupils got near to the median mark.

4 a The sample needs to be representative of people living on the street: male/female, adults/children.

   b 120 people

Number
Integers, decimals and symbols

NAILIT!

Make sure you understand terms such as integer, place value, ascending and descending.

(1) Arrange these numbers in descending order. (2 marks, ★)

$(-1)^3$ $(0.1)^2$ $\frac{1}{1000}$ 0.1 $\frac{1}{0.01}$

....

(2) Without using a calculator, work out (★)

a 0.035×1000 (1 mark) ....

b $12.85 \div 1000$ (1 mark) ....

c $(-3) \times 0.09 \times 1000$ (1 mark) ....

d $(-1) \times (-0.4) \times 100$ (1 mark) ....

[Total: 4 marks]

(3) $0.86 \times 54 = 46.44$

Without using a calculator, work out (★★) ◄—————— Think about place value.

a 86×54 (1 mark)   **c** $\frac{4644}{54}$ (1 mark)

....

b 8.6×540 (1 mark)   **d** $\frac{46.44}{0.086}$ (1 mark)

....

[Total: 4 marks]

(4) Place the correct symbol from the following list in the box.
The symbols can be used once, more than once or not at all. (★★)

$<$ $\geq$ $\leq$ $=$ $>$

a 12.56×3.45 ☐ 0.1256×345 (1 mark) ◄——— Work out the calculation either side of the box and then insert the correct symbol in the box.

b $(-8)^2$ ☐ -64 (1 mark)

c $6 - 12$ ☐ $8 - 14$ (1 mark)

d $(-7) \times (0)$ ☐ $(-7) \times (-3)$ (1 mark)

[Total: 4 marks]

Addition, subtraction, multiplication and division

Do **not** use a calculator for any of these calculations.

NAILIT!

On Paper 1 these calculations must be done without a calculator.

(1) Work out (★★)

a 67.78 + 8.985 (1 mark)

c 93.1 − 1.77 (1 mark)

....

....

WORKIT!

Work out 23.48 − 8.362.

$$\begin{array}{r} {}^{1}2\,{}^{1}3\,.\,4\,{}^{7}\!\not{8}\,{}^{1}0 \\ -\quad 8\,.\,3\,\;6\,\;2 \\ \hline 1\,5\,.\,1\,\;1\,\;8 \end{array}$$

b 124.706 + 76.9 + 0.04
(1 mark)

d 23.7 + 8.94 − 22.076
(1 mark)

....

....

[Total: 4 marks]

(2) Work out (★★)

a 147 × 8 (1 mark)

c 9.7 × 4.6 (1 mark)

e 486 ÷ 18 (1 mark)

....

....

....

WORKIT!

Work out 8.97 ÷ 1.3.

Multiply the numerator and denominator by a factor of 10 to make the denominator a whole number.

b 57 × 38 (1 mark)

d 1.24 × 0.53 (1 mark)

f 94.5 ÷ 1.5 (1 mark)

$$\frac{8.97}{1.3} = \frac{89.7}{13}$$

$$\begin{array}{r} 6\,.\,9 \\ 13\overline{)8\;9\,.\,7} \\ 7\;8 \\ \hline 1\;1\;7 \\ 1\;1\;7 \\ \hline 0 \end{array}$$

....

....

....

[Total: 6 marks]

(3) Work out (★★★)

a 34^2 (1 mark)

b $\dfrac{1.5 \times 2.5}{0.5}$ (1 mark)

c 2.4^2 (1 mark)

....

....

....

[Total: 3 marks]

Using fractions

(1) Fill in the missing numbers in these equivalent fractions. (1 mark, ★)

$$\frac{2}{5} = \frac{16}{\Box} = \frac{\Box}{75} = \frac{50}{\Box}$$

(2) Simplify these fractions. Give your answers as mixed numbers in their simplest form. (★★)

a $\frac{64}{12}$ (1 mark) ....

b $\frac{124}{13}$ (1 mark) ....

[Total: 2 marks]

(3) Work out these calculations, simplifying your answer if possible. (★★★)

a $4\frac{1}{4} \times 1\frac{2}{3}$ (2 marks)

....

b $1\frac{7}{8} \div \frac{1}{4}$ (2 marks)

....

c $3\frac{1}{5} - \frac{3}{4}$ (2 marks)

....

[Total: 6 marks]

(4) In a class of students, all the students walk, cycle, travel by bus or travel by car to school. $\frac{2}{7}$ of the students walk, $\frac{3}{8}$ of the students cycle and $\frac{1}{4}$ travel by bus. What fraction of the students travel to school by car? (2 marks, ★★★)

....

(5) Put the following list of fractions into order of size starting with the smallest. (2 marks, ★★)

$$\frac{2}{3} \qquad \frac{3}{4} \qquad \frac{7}{8} \qquad \frac{1}{2} \qquad \frac{7}{12}$$

....

Different types of number

(1) Use the numbers from the list to answer the following questions. (★)

2   7   49   6   14

Write down the number that is

a   a factor of 77 (1 mark)

....

b   a square number (1 mark)

....

c   an even prime number (1 mark)

....

d   not a factor of 98 (1 mark)

....

e   a multiple of 3. (1 mark)

....

NAILIT!

Make sure you understand terms such as multiple, factor, common factor, highest common factor, lowest common multiple and prime.

STRETCHIT!

A **perfect number** is one whose factors add up to the number itself. An **abundant number** has factors that add up to more than the number itself. Write a factor tree for the number 216. Is it one of these?

[Total: 5 marks]

WORKIT!

The number 540 can be written as the product of its prime factors:

$540 = 2^2 \times 3^3 \times 5$

a   Write 168 as the product of its prime factors.

$168 = 2 \times 84$

$\quad\quad = 2 \times 2 \times 42$ ←

All the numbers in the product must be prime. If they aren't, keep breaking the number down until they are.

$\quad\quad = 2 \times 2 \times 6 \times 7$

$\quad\quad = 2 \times 2 \times 2 \times 3 \times 7$

$\quad\quad = 2^3 \times 3 \times 7$

b   Work out the highest common factor of 540 and 168.

Both numbers have 2^2 and 3 in their prime factors,   ← $2^3 = 2^2 \times 2$
so highest common factor $= 2^2 \times 3 = 12$

c   Work out the lowest common multiple of 540 and 168.

Multiplying the prime factors of both, ignoring duplicates,
the lowest common multiple $= 2^3 \times 3^3 \times 5 \times 7 = 7560$

(2) The number 945 can be written as the product of prime factors: $945 = 3^3 \times 5 \times 7$. (★★)

    a Write the number 693 as the product of prime factors. (2 marks)

....

    b Work out the highest common factor of 945 and 693. (2 marks)

....

    c Work out the lowest common multiple of 945 and 693. (2 marks)

....

[Total: 6 marks]

(3) Find the lowest common multiple of 49 and 63. (2 marks, ★★)

....

(4) Tom rings a bell every 15 seconds, Julie bangs a drum every 20 seconds and Phil hits a triangle every 25 seconds. If they all sound a note together at the same time, after how many minutes will they next make a note at the same time? (2 marks, ★★★)

> Start by finding the product of prime factors for each time. Then find the lowest common multiple.

.... minutes

Listing strategies

NAILIT!

Sometimes it's hard to spot a method for solving a question. Making a list can make it easier to see what to do next.

(1) A frog croaks every 14 seconds and a parrot squawks every 15 seconds.

Assuming that they both make their sound together initially, what is the minimum amount of time in seconds before they will sound together again? (2 marks, ★★★)

> Write down all the multiples of the times to find the first time which is common to both lists.

.... seconds

(2) Dhaya has some chocolates which she shares with her friends.

If she gives 6 chocolates to each friend she will have one chocolate left.

If she gave each friend 7 sweets she would need another 4 sweets.

> Start by listing the multiples of 6 and 7.

By producing lists, find the number of friends she could have. (3 marks, ★★★)

....

(3) The ratio of male to female students in a college is 5:6.
There are 100 more female students than male students in the college.
Find the total number of students in the college. (3 marks, ★★★)

....

(4) A group of 6 students in a primary school are asked to work in pairs on a science project.
How many possible pairs could there be? (3 marks, ★★★)

> Represent each pupil by a letter and write down the possible pairs excluding any pairs that are the same.

....

219

The order of operations in calculations

Do **not** use a calculator for any of these calculations.

(1) Ravi is finding the value of this expression. (★)

$24 \div 3 + 8 \times 4$. ◄─────────

Use BIDMAS: brackets, indices (powers and roots), division, multiplication, addition, subtraction.

Here is his working out.

$$24 \div 3 + 8 \times 4 = 8 + 8 \times 4$$
$$= 16 \times 4$$
$$= 64$$

Ravi's answer is wrong.

NAILIT!

Do not work out calculations in the order they appear from left to right. Use BIDMAS.

a Explain what he has done wrong. (3 marks)

....

....

b Showing your working, what is the correct answer? (1 mark)

....

[Total: 4 marks]

(2) Evaluate (★★)

a $9 \times 7 \times 2 - 24 \div 6$ (1 mark)

....

b $2 - (-27) \div (-3) + 4$ (1 mark)

....

c $(4 - 16)^2 \div 4 + 32 \div 8$ (1 mark)

....

[Total: 3 marks]

(3) Evaluate (★★★)

a $1 + 4 \div \dfrac{1}{2} - 3$ (1 mark)

....

b $15 - (1 - 2)^2$ (1 mark)

....

c $\sqrt{4 \times 12 - 2 \times (-8)}$ (1 mark)

....

[Total: 3 marks]

Indices

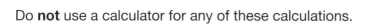

Do **not** use a calculator for any of these calculations.

(1) Express each of these as a single power of 10. (★★)

   a $10^5 \times 10$ (1 mark)

   c $\dfrac{10^5 \times 10^3}{10^2}$ (1 mark)

   ....

   ....

   b $(10^4)^2$ (1 mark)

   d $(10^6)^{\frac{1}{2}}$ (1 mark)

   ....

   ....

NAILIT!

Make sure that you learn the rules of indices.

NAILIT!

If a number appears on its own without a power then the power to which it is raised is 1. So 10 can be written as 10^1.

[Total: 4 marks]

(2) Without using a calculator, write the value of (★★★)

   a 5^0 (1 mark)

   c $8^{\frac{1}{3}}$ (1 mark)

   ....

   ....

   b 3^{-2} (1 mark)

   d $49^{\frac{1}{2}}$ (1 mark)

   ....

   ....

[Total: 4 marks]

(3) Without using a calculator, work out (★★★★)

   a $\left(\dfrac{8}{27}\right)^{-\frac{1}{3}}$ (2 marks)

   c $36^{-\frac{1}{2}}$ (2 marks)

   ....

   ....

   b $\left(\dfrac{1}{4}\right)^{-2}$ (2 marks)

   d $16^{\frac{3}{2}}$ (2 marks)

   ....

   ....

[Total: 8 marks]

(4) Find the value of x such that $5^{2x} = 125$. (3 marks, ★★★★)

   ....

Surds

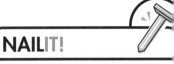

(1) Evaluate (★★★★)

   a $\left(\sqrt{5}\right)^2$ (1 mark)   **b** $3\sqrt{2} \times 5\sqrt{2}$ (1 mark)   **c** $\left(3\sqrt{2}\right)^2$ (2 marks)

> **NAILIT!**
>
> The square of a square root is the number inside the square root, e.g. $\left(\sqrt{2}\right)^2 = 2$.

....

[Total: 4 marks]

(2) Rationalise the denominator and simplify $\dfrac{15}{4\sqrt{3}}$. (2 marks, ★★★★)

> **NAILIT!**
>
> 'Rationalise' means to remove the surd from the denominator. To do this you multiply both numerator and denominator by that surd.

....

(3) Show that $(2 + \sqrt{3})(2 - \sqrt{3})$ simplifies to 1. (2 marks, ★★★★)

(4) $\sqrt{3}\left(3\sqrt{24} + 4\sqrt{6}\right)$ can be simplified to $a\sqrt{2}$. Find the value of a. (3 marks, ★★★★★)

> Multiply every term in the bracket by the term outside the bracket.

....

(5) Work out $(1 - \sqrt{5})(3 + 2\sqrt{5})$. (2 marks, ★★★★)

> Multiply every term in the first bracket by every term in the second bracket.

....

(6) Show that $\dfrac{1}{\sqrt{2}} + \dfrac{1}{4}$ can be written as $\dfrac{1 + 2\sqrt{2}}{4}$. (4 marks, ★★★★)

> To remove the surd in, for example, $\dfrac{5}{1-\sqrt{3}}$ multiply the numerator and denominator by $1 + \sqrt{3}$.

(7) Show that $\dfrac{2}{1-\frac{1}{\sqrt{2}}}$ can be written as $4 + 2\sqrt{2}$. (4 marks, ★★★★★)

(8) Show that $\dfrac{3}{\sqrt{3}} + \sqrt{75} + (\sqrt{2} \times \sqrt{6}) = 8\sqrt{3}$. (3 marks, ★★★★★)

Standard form

SNAPIT! **Standard form**

Numbers in standard form are in the form $a \times 10^n$ where $1 \le a < 10$ and n is an integer (positive, negative or zero).

(1) Write these numbers in standard form. (★)

a 0.00255 (1 mark)

c 0.000000089 (1 mark)

....

....

b 10060 000 000 (1 mark)

....

[Total: 3 marks]

(2) Without using a calculator, work out these calculations. Give your final answer in standard form. (★★★★)

a $3 \times 10^6 \times 2 \times 10^8$ (1 mark) ◄── Multiply the number parts. Multiply the powers of 10 using the index laws.

....

b $5.5 \times 10^{-3} \times 2 \times 10^8$ (2 marks)

....

c $\dfrac{8 \times 10^5}{4 \times 10^3}$ (1 mark)

....

d $\dfrac{5 \times 10^{-3}}{0.5}$ (2 marks)

....

e $\dfrac{4.5 \times 10^{-6}}{0.5 \times 10^{-3}}$ (1 mark) ◄── If necessary, change the decimal point in the number and the power of 10 to give the answer in standard form.

....

WORKIT!

Without using a calculator, work out, giving your final answer in standard form.

a $3.5 \times 10^{-4} \times 4 \times 10^8$

$3.5 \times 10^{-4} \times 4 \times 10^8$

$= (3.5 \times 4) \times (10^{-4} \times 10^8)$

$= 14 \times 10^4$

$= 1.4 \times 10^5$

b $(4.5 \times 10^{-2}) \div (6 \times 10^{-5})$

$(4.5 \times 10^{-2}) \div (6 \times 10^{-5})$

$= (4.5 \div 6) \times (10^{-2} \div 10^{-5})$

$= 0.75 \times 10^3$

$= 7.5 \times 10^2$

[Total: 7 marks]

(3) Without using a calculator, work out $2.4 \times 10^3 + 2.8 \times 10^2$.

Give your answer as an ordinary number. (2 marks, ★★★)

....

(4) If $a \times 10^4 + a \times 10^2 = 33330$, find the value of a. (3 marks, ★★★★)

....

Converting between fractions and decimals

(1) Without using a calculator, convert these fractions into decimals. (★★)

a $\frac{11}{20}$ (1 mark)

b $\frac{3}{8}$ (1 mark)

....

....

NAILIT!

Check that you know all the common fractions as decimal conversions (e.g. $\frac{1}{5} = 0.2$).

[Total: 2 marks]

(2) Without using a calculator, state whether each of these fractions will produce a terminating or a recurring decimal. (★★★)

a $\frac{1}{16}$ (1 mark)

....

b $\frac{5}{7}$ (1 mark)

....

c $\frac{13}{35}$ (1 mark)

....

STRETCHIT!

Look at the prime factors in the denominator. If they are all 2 and/or 5, then the decimal will terminate.

[Total: 3 marks]

(3) Prove that the recurring decimal $0.\dot{4}0\dot{2}$ can be written as the fraction $\frac{134}{333}$. (3 marks, ★★★★)

Start by letting x be equal to the recurring decimal.

WORKIT!

Write $0.\dot{7}\dot{1}$ as a fraction.

$$\text{Let } x = 0.717171...$$
$$100x = 71.717171...$$
$$100x - x = 71$$
$$99x = 71$$
$$x = \frac{71}{99}$$

(4) Write the recurring decimal $0.6\dot{5}\dot{2}$ as a fraction in its simplest form. (3 marks, ★★★★)

Remember to cancel the resulting fraction.

....

Converting between fractions and percentages

(1) Convert these percentages to fractions in their simplest form. (★)

a 35% (1 mark)

c 76% (1 mark)

....

....

b 7% (1 mark)

d 12.5% (1 mark)

← Write the percentage as a fraction with denominator 100. Simplify as far as possible.

....

....

[Total: 4 marks]

(2) Without using a calculator, write each of these fractions as percentages. (★★)

a $\frac{1}{5}$ (1 mark)

c $\frac{150}{60}$ (1 mark)

....

....

Where the denominator is a factor of 100, multiply both the top and the bottom of the fraction by the amount needed to get 100 on the bottom. ←

b $\frac{17}{25}$ (1 mark)

d $\frac{7}{40}$ (1 mark)

....

....

[Total: 4 marks]

(3) Convert the fraction $\frac{8}{15}$ to a percentage. Give your answer to 2 decimal places. (2 marks, ★★)

....

(4) Jake scored $\frac{66}{90}$ in a physics test and 75% in a chemistry test. Show, with reasons, which test he did better in. (2 marks, ★★★)

NAILIT!

Where the denominator is not a factor of 100, divide the numerator by the denominator and multiply by 100.

Fractions and percentages as operators

(1) Calculate 70% of £49.70. (1 mark, ★)

....

(2) In a batch of apples, 8% of the apples are bad. If there are 600 apples in the batch, how many of them are bad? (2 marks, ★★)

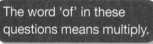

The word 'of' in these questions means multiply.

....

(3) In a batch of car parts, 12% of them were rejected. How many car parts are likely to be accepted in a batch of 8000? (2 marks, ★★)

....

WORKIT!

1 Find $\frac{2}{3}$ of £120.

$$\frac{120}{3} \times 2$$

$$= 40 \times 2 = £80$$

2 Find 18% of 50 kg.

$$0.18 \times 50 = 9\,kg$$

(4) Chloe buys a new car costing £12 000 + VAT at 20%. (★★★)

a Calculate the total cost of the car including VAT. (1 mark)

....

b Chloe has to pay the garage 20% of the total cost of the car as a deposit and the remainder is paid over 36 equal monthly payments. Calculate her monthly payment. (2 marks)

....

[Total: 3 marks]

(5) In a class of students, $\frac{2}{3}$ are male. Out of the males $\frac{4}{11}$ are taking chemistry.

What fraction of the class are males **not** taking chemistry? (2 marks, ★★★)

....

Standard measurement units

(1) Convert 1.75 km to centimetres. (2 marks, ★)

.... cm

(2) A carton contains 3 litres of juice. How many complete 175 ml glasses could be filled from the carton? (3 marks, ★★)

....

(3) A children's paddling pool is to be filled with 900 litres of water. The only container they can find to fill the paddling pool has a volume of 700 cm³.

How many containers full of water would be required to fill the paddling pool? Give your answer to the nearest whole number. (3 marks, ★★)

....

(4) 12 g of carbon contains 6.02×10^{23} atoms. (★★★★)

a Find the mass of 1 atom of carbon. Give your answer in standard form in g, correct to 3 significant figures. (2 marks)

.... g

b Give your answer to part a in kg, also in standard form and correct to 3 significant figures. (2 marks)

.... kg

[Total: 4 marks]

(5) An electron has a mass of 9.11×10^{-31} kg. In an atom of gold there are 79 electrons. Work out the mass of electrons in one atom of gold. Give your answer in standard form to 3 significant figures in grams. (3 marks, ★★★★)

.... g

NAILIT!

Learn the conversion factors for units of length, mass, capacity and time.

NAILIT!

Always ask yourself before you start 'is the answer going to be bigger or smaller than the number in the question?'.

First change the volumes into the same units.

Rounding numbers

(1) Write these numbers correct to the nearest whole number. (★)

a 34.7 (1 mark)

....

b 100.6 (1 mark)

....

c 0.078 (1 mark)

....

d 0.4999 (1 mark)

....

e 1.5001 (1 mark)

....

WORKIT!

Round 6.785 to 2 decimal places.

1 Look at the 3rd digit after the decimal place: 5.

2 If is 5 or more, round up; if it is less than 5, round down.

6.785 = 6.79 (to 2 d.p.)

[Total: 5 marks]

(2) The number 34.8765 is given to 4 decimal places. (★)

Write this number to

a 2 decimal places (1 mark)   **b** 3 decimal places. (1 mark)

....

[Total: 2 marks]

(3) Write each number to the specified number of significant figures. (★)

a 12758 (3 s.f.) (1 mark)

....

b 0.01055 (2 s.f.) (1 mark)

....

c 7.46 × 10⁻⁵ (1 s.f.) (1 mark)

....

WORKIT!

Round 0.078366 to 3 significant figures.

1 Identify the first significant figure: 7.

2 Count the required number of figures: 0.0783.

3 Round the next digit: 0.0784.

0.078366 = 0.0784 (to 3 s.f.)

[Total: 3 marks]

(4) Find the value of these calculations. Give all your answers to 3 significant figures. (★★)

a $(0.18 \times 0.046)^2 - 0.01$ (1 mark)

....

b $\dfrac{1200 \times 1.865}{2.6 \times 25}$ (1 mark)

....

c $\dfrac{36}{0.07} \times 12 \div \dfrac{1}{2}$ (1 mark)

....

[Total: 3 marks]

Estimation

(1) Estimate the value of $4.6 \times 9.8 \times 3.1$. Give your answer to 1 significant figure. (2 marks, ★)

....

(2) **a** Work out $19.87^2 - \sqrt{404} \times 7.89$.

 Write down all the numbers on your calculator display. (2 marks, ★)

....

b Use estimation to check your answer to part a, showing all your working. (1 mark, ★)

....

[Total: 3 marks]

(3) Work out an estimate for $(0.52 \times 0.83)^2$. (2 marks, ★★)

WORKIT!

Estimate the value of $\sqrt{31}$.

1 Find the square root of the perfect squares that are above and below the value.

$\sqrt{25} = 5$ and $\sqrt{36} = 6$.

2 Take the value between them and square it to see whether this makes a good estimate.

5.5^2 is 30.25, just slightly less than 31. So 5.5 is a reasonable estimate.

....

(4) Work out an estimate for $\sqrt{5.08} + 4.10 \times 5.45$ (2 marks, ★★★★)

....

(5) Estimate the value of $\sqrt{112}$. Give your answer to 1 decimal place. (2 marks, ★★★) ◀——— 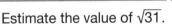 Write the numbers above and below 112 that are perfect squares.

....

(6) Estimate the value of

$$\frac{0.89 \times 7.51 \times 19.76}{2.08 \times 5.44 \times 3.78}$$

Give your answer to 1 significant figure. (3 marks, ★★)

....

(7) An electron has a mass of 9.11×10^{-31} kg. An atom of carbon contains 6 electrons. (★★★★★)

a Work out an estimate for the mass of the electrons in 100 carbon atoms. Give your answer correct to 1 significant figure in standard form in kg. (2 marks)

....kg

b Is your answer to part a an overestimate or underestimate? Give a reason for your answer. (1 mark)

....

....

....

....

[Total: 3 marks]

Upper and lower bounds

① The mass of a brick, m, is 2.34 kg correct to 3 significant figures.

Write down the error interval for the mass of the brick. (2 marks, ★★)

....

② The power, P, in watts produced in an electric resistor is given by the formula

$P = IV$

where I is the current in amps and V is the voltage in volts.

$P = 3.052$ correct to 4 significant figures

$I = 1.24$ amps correct to 3 significant figures (★★★★★)

a i Calculate the upper bound for V. (1 mark)

....

ii Calculate the lower bound for V. (1 mark)

....

....

b Using your answers for part a, work out the value of V to a suitable degree of accuracy. Give a reason for your answer. (2 marks)

....

[Total: 4 marks]

③ A bookcase has a shelf length of 1.1 m correct to 2 significant figures.

A maths textbook is 3.0 cm thick correct to 2 significant figures.

How many books will definitely fit onto the shelf? (3 marks, ★★★★★)

....

Algebra
Simple algebraic techniques

NAILIT!

Make sure you understand the difference between terms such as equation, expression, identity and formula.

(1) Identify whether each of these is a formula, expression, equation or identity. (★)

  a $v^2 = u^2 + 2as$ (1 mark)

....

  d $(2a^2b)^2 = 4a^4b^2$ (1 mark)

....

  b $5x(2x + y) = 10x^2 + 5xy$ (1 mark)

....

  e $P = I^2R$ (1 mark)

....

  c $6a^2b$ (1 mark)

....

[Total: 5 marks]

NAILIT!

Collect together like terms, with identical letters and powers.

(2) Simplify $4x + 3x \times 2x - 3x$. (2 marks, ★★)

....

(3) Karl is trying to work out two values of y for which $y^3 - y = 0$.

The two values he finds are 1 and -1.

Are these two values correct? You must show your working. (3 marks, ★★★)

....

(4) Simplify these expressions. (★★★)

  a $6x - (-4x)$

     (1 mark)

  b $x^2 - 2x - 4x + 3x^2$

     (1 mark)

  c $(-2x)^2 + 6x \times 3x - 4x^2$

     (2 marks)

....

[Total: 4 marks]

(5) $s = \dfrac{v^2 - u^2}{2a}$ (★★★)

Work out the value of s when

  a $v = 3$, $u = 1$ and $a = 2$

     (1 mark)

  b $v = -4$, $u = 3$ and $a = 4$

     (1 mark)

  c $v = 5$, $u = -2$ and $a = -7$

     (1 mark)

....

[Total: 3 marks]

Removing brackets

(1) Expand the brackets for these expressions. (★)

   a $8(3x - 7)$ (1 mark)   **b** $-3(2x - 4)$ (1 mark)

.... **[Total: 2 marks]**

NAILIT!

Watch out for negative numbers outside the bracket as the signs will change when you multiply them out.

(2) Simplify (★★)

   a $3(2x - 1) - 3(x - 4)$ (2 marks)   **c** $5ab(2a - b)$ (2 marks)

....

   b $4y(2x + 1) + 6(x - y)$ (2 marks) **d** $x^2y^3(2x + 3y)$ (2 marks)

NAILIT!

Use the laws of indices when you expand brackets.

.... **[Total: 8 marks]**

(3) Expand and simplify (★★★★)

   a $(m - 3)(m + 8)$ (2 marks)   **c** $(3x - 1)^2$ (2 marks)

....

   b $(4x - 1)(2x + 7)$ (2 marks)   **d** $(2x + y)(3x - y)$ (3 marks)

.... **[Total: 9 marks]**

(4) Expand and simplify (★★★★)

   a $(x + 5)(x + 2)$ (2 marks)   **c** $(x - 7)(x + 1)$ (2 marks)

....

   b $(x + 4)(x - 4)$ (2 marks)   **d** $(3x + 1)(5x + 3)$ (2 marks)

.... **[Total: 8 marks]**

(5) Expand and simplify (★★★★★)

   a $(x + 3)(x - 1)(x + 4)$ (3 marks)   **b** $(3x - 4)(2x - 5)(3x + 1)$ (3 marks)

.... **[Total: 6 marks]**

Factorising

(1) Factorise fully (★★★)

  a $25x^2 - 5xy$ (2 marks)
      b $4\pi r^2 + 6\pi x$ (2 marks)
      c $6a^3b^2 + 12ab^2$ (2 marks)

....

[Total: 6 marks]

WORKIT!

Factorise $15xy + 3x^2$.

> Take the common factors outside the brackets.

$15xy + 3x^2 = 3x(5y + x)$

SNAPIT! Factorising

Factorising is the reverse process to expanding the brackets.

(2) Factorise (★★★)

  a $9x^2 - 1$ (2 marks)

....

  b $16x^2 - 4$ (2 marks)

....

> Use the difference of two squares:
> $a^2 - b^2 = (a + b)(a - b)$

[Total: 4 marks]

NAILIT!

Make sure that you take out all factors.

When a question says 'factorise fully', there is usually more than one factor. But if the question just says 'factorise' still check for more than one factor.

(3) Factorise (★★★)

  a $a^2 + 12a + 32$ (2 marks)
      b $p^2 - 10p + 24$ (2 marks)

....

[Total: 4 marks]

WORKIT!

Factorise $x^2 - 2x - 8$.

> Find two numbers that multiply to make -8 and sum to make -2.

2 and -4

$x^2 - 2x - 8 = (x + 2)(x - 4)$

(4) Factorise (★★★)

  a $a^2 + 12a$ (2 marks)
      c $x^2 - 11x + 30$ (2 marks)

....

  b $b^2 - 9$ (2 marks)

.... **[Total: 6 marks]**

(5) Factorise (★★★★★)

a $3x^2 + 20x + 32$ (3 marks)

c $2x^2 - x - 10$ (3 marks)

....

....

b $3x^2 + 10x - 13$ (3 marks)

....

[Total: 9 marks]

(6) Work out $\dfrac{x + 15}{2x^2 - 3x - 9} + \dfrac{3}{2x + 3}$.

Give your answer in its simplest form. (4 marks, ★★★★★)

....

(7) Write $\dfrac{1}{8x^2 - 2x - 1} \div \dfrac{1}{4x^2 - 4x + 1}$ in the form $\dfrac{ax + b}{cx + d}$ where a, b, c and d are integers.

(3 marks, ★★★★★)

....

Changing the subject of a formula

(1) Make T the subject of the formula $PV = nRT$. (2 marks, ★★)

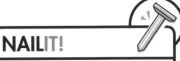
....

(2) Make y the subject of the formula $2y + 4x - 1 = 0$. (2 marks, ★★★)

....

(3) Make a the subject of the formula $v = u + at$. (2 marks, ★★★)

....

(4) Make x the subject of the formula $y = \frac{x}{5} - m$. (2 marks, ★★★)

....

(5) Make v the subject of the formula $E = \frac{1}{2}mv^2$. (2 marks, ★★★)

....

(6) The volume of a cone is given by the formula $V = \frac{1}{3}\pi r^2 h$ where V is the volume, r is the radius and h is the perpendicular height. (★★★)

  a Rearrange the formula to make r the subject. (2 marks)

....

  b Find the radius of a cone with a volume of $100\,\text{cm}^3$ and a height of $8\,\text{cm}$. Give your answer to 2 decimal places. (2 marks)

....

[Total: 4 marks]

(7) A straight line has the equation $y = 3x - 9$. (★★★)

    a Rearrange the equation to make x the subject. (1 mark)

....

    b Find the value of x when $y = 3$. (1 mark)

....

[Total: 2 marks]

(8) Make x the subject of $3y - x = ax + 2$. (4 marks, ★★★★)

NAILIT!

When the subject appears on both sides, first get the terms containing the subject on one side. Then collect like terms or factorise to make sure the subject only appears once.

....

(9)   **a** Make c the subject of the formula $c^2 = \dfrac{(16a^2\, b^4\, c^2)^{\frac{1}{2}}}{4a^2\, b}$.

    (2 marks, ★★★★★)

....

    b $a = 2.8$ and $b = 3.2$, both to 1 decimal place. (3 marks, ★★★★★)

    Work out the upper and lower bounds of c. Give your answers to 3 significant figures.

....

[Total: 5 marks]

Solving linear equations

(1) Solve these equations. (★★)

a $2x + 11 = 25$ (1 mark)

e $\dfrac{4x}{5} = 20$ (1 mark)

....

....

b $3x - 5 = 10$ (1 mark)

f $\dfrac{2x}{3} = -6$ (1 mark)

....

....

c $15x = 60$ (1 mark)

g $5 - x = 7$ (1 mark)

....

....

d $\dfrac{x}{4} = 8$ (1 mark)

h $\dfrac{x}{7} - 9 = 3$ (1 mark)

....

....

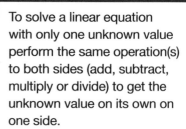

NAILIT!

To solve a linear equation with only one unknown value perform the same operation(s) to both sides (add, subtract, multiply or divide) to get the unknown value on its own on one side.

[Total: 8 marks]

(2) Solve the equation $5x - 1 = 2x + 1$. (2 marks, ★★)

....

(3) Solve the equations. (★★★)

a $\dfrac{1}{4}(2x - 1) = 3(2x - 1)$ (3 marks)

....

....

b $5(3x + 1) = 2(5x - 3) + 3$ (3 marks)

....

[Total: 6 marks]

Solving quadratic equations using factorisation

WORKIT!

Solve $x^2 - 2x - 2 = 2x + 3$.

1 Rewrite the equation as a quadratic equal to zero.

$x^2 - 4x - 5 = 0$

2 Factorise.

$(x - 5)(x + 1) = 0$

3 Set each bracket equal to zero.

$x - 5 = 0$ or $x + 1 = 0$

$x = 5$ or $x = -1$

NAILIT!

Make sure you can expand brackets and factorise before attempting these questions.

NAILIT!

Remember that a quadratic must be equal to zero before you can factorise and hence solve it.

---

(1) **a** Factorise $x^2 - 7x + 12$. (2 marks, ★★★)

....

b Solve $x^2 - 7x + 12 = 0$. (1 mark, ★★★)

....

[Total: 3 marks]

(2) **a** Factorise $2x^2 + 5x - 3$. **b** Solve $2x^2 + 5x - 3 = 0$.
   (3 marks, ★★★★)   (1 mark, ★★★)

NAILIT!

You need to recognise quadratic expressions that can be factorised into two brackets.

....

[Total: 4 marks]

(3) Solve the equation $x^2 - 3x - 20 = x - 8$. (4 marks, ★★★★)

....

④ **a** Show that the equation $x(x - 8) - 7 = x(5 - x)$ can be rearranged to
$2x^2 - 13x - 7 = 0$. (2 marks, ★★★★)

NAILIT!

If a quadratic equation appears in a problem, always check whether both solutions are possible or only one of them.

b Hence find the solutions to $x(x - 8) + 7 = x(5 - x)$. (3 marks, ★★★★)

....

[Total: 5 marks]

⑤ The diagram shows a trapezium with the sides measured in cm.

The trapezium has an area of $16\,\text{cm}^2$.

Find the value of x. (4 marks, ★★★★★)

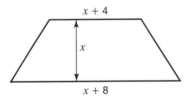

NAILIT!

Area of a trapezium
$= \frac{1}{2}(a + b)h$
where a and b are the lengths of the two parallel sides and h is the distance between them.

....

Solving quadratic equations using the formula

(1) **a** Show that $\frac{3}{x+7} = \frac{2-x}{x+1}$ can be written as $x^2 + 8x - 11 = 0$.

(3 marks, ★★★★★)

....

b Hence solve the equation $\frac{3}{x+7} = \frac{2-x}{x+1}$.

Give your answers to 2 decimal places. (2 marks, ★★★★★)

NAILIT!

Quadratic equations of the form $ax^2 + bx + c = 0$ $(a \neq 0)$ can be solved using the formula

$$x = \frac{-b + \sqrt{b^2 - 4ac}}{2a}$$

NAILIT!

Be careful with the signs when entering numbers into the formula. If you end up with the square root of a negative number you have made a mistake.

[Total: 5 marks]

(2) The right-angled triangle shown has an area of $40\,\text{cm}^2$.
All measurements are in centimetres. (5 marks, ★★★★★)

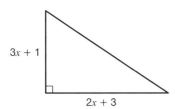

3x + 1

2x + 3

Find the value of x to 2 decimal places.

....

(3) Calculate the x-coordinates of the points of intersection of the line $y = x - 8$ and the curve $y = x^2 - 2x - 9$. Give your answers to 2 decimal places. (5 marks, ★★★★★)

NAILIT!

If you are asked to solve a quadratic equation giving your answer to a certain number of decimal places then you probably need to use the formula.

....

Solving simultaneous equations

① Solve these simultaneous equations algebraically.

$2x - 3y = -5$

$5x + 2y = 16$ (3 marks, ★★★)

> The opposite signs for the y terms mean that it is easier to make these terms the same in value but opposite in sign and add the two equations.

NAILIT!

You can solve simultaneous equations:
- graphically, by finding where the equations intersect on a graph
- by eliminating one of the unknowns by adding or subtracting the equations
- by substituting the expression for one variable into the other equation.

NAILIT!

Be careful with the signs when solving simultaneous equations using the elimination method, especially if you have to subtract one equation from the other.

$x = $ .... $y = $ ....

② a Plot the graph of $y = 3x - 2$ on the set of axes given. (2 marks, ★★★)

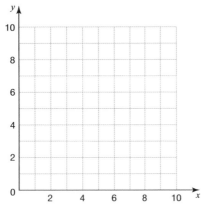

NAILIT!

Only use x values that are on the graph and check that the resulting y values will fit on the graph. You only need two points to plot a straight line, but plotting three points means that you can identify whether you have made a mistake.

b Hence solve the pair of simultaneous equations (3 marks, ★★★)

$y = 3x - 2$

$y = 10 - x$

$x = $ .... $y = $ ....

[Total: 5 marks]

③ Solve these simultaneous equations.

$x - y = 3$

$x^2 + y^2 = 9$ (5 marks, ★★★★★)

NAILIT!

If one of the equations is a circle or a quadratic, you must use the substitution method or draw the graphs.

$x = $ .... $y = $ ....

Solving inequalities

1. Solve these inequalities. (★★★)

   a $\dfrac{x+5}{4} \geq -1$

   (1 mark)

   b $3x - 4 > 4x + 8$

   (2 marks)

....

....

[Total: 3 marks]

2. Show the inequality $-3 < x \leq 2$ on the number line below. (2 marks, ★★)

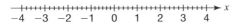

3. a Use the grid to shade the region represented by these inequalities. (5 marks, ★★★★★)

   $x \leq 1 \quad y > -2 \quad y - 2x < 1$

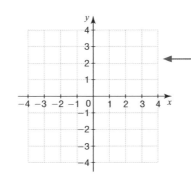

   Show < or > with a dotted line, and ≤ or ≥ with a solid line. Shade inside the region enclosed by the lines.

   b List the integer values of coordinates that satisfy all of these inequalities. (2 marks, ★★★★★)

   ....

   [Total: 7 marks]

4. Solve the inequality $x^2 + 2x \leq 3$. (5 marks, ★★★★★)

   ....

5. Solve the inequality $x^2 - 2x - 15 > 0$. (5 marks, ★★★★★)

   ....

Problem solving using algebra

(1) The length of a rectangular patio is 1 m more than its width.

The perimeter of the patio is 26 m. Find the area of the patio. (3 marks, ★★★)

NAILIT!

Look for all the algebraic techniques when solving problems. You will often need to substitute letters for unknown values.

....

(2) The cost of 2 adults' tickets and 5 children's tickets at a circus is £35.

The cost of 3 adults' tickets and 4 children's tickets is £38.50.

Find the cost of each type of ticket. (3 marks, ★★★★)

....

(3) Rachel and Hannah are sisters.

The product of their ages is 63. In two years' time, the product of their ages will be 99. (★★★★★)

a Find the sum of their ages. (3 marks)

....

b Rachel is 2 years older than Hannah. How old is Rachel? (2 marks)

....

[Total: 5 marks]

Use of functions

(1) $f(x) = 5x + 4$ (★★)

    a Find f(3). (1 mark)

....

    b Find the value of x for which $f(x) = -1$. (2 marks)

....

                                                         [Total: 3 marks]

(2) If $f(x) = x^2$ and $g(x) = x - 6$, find (★★★)

    a fg(x) (1 mark)   **b** gf(x) (2 marks)

....

                                                             [Total: 3 marks]

(3) $f(x) = \sqrt{x + 4}$, where $x > -4$

    $g(x) = 2x^2 - 3$ for all values of x (★★★★)

    a Find f(5). (1 mark)   **b** Find an expression for gf(x).
                                                Simplify your answer. (2 marks)

....

                                                             [Total: 3 marks]

(4) $f(x) = 5x^2 + 3$

    Find $f^{-1}(x)$. (3 marks, ★★★★★)

NAILIT!

If $f(x) = x^2 - 1$,
to find f(1) substitute $x = 1$:
$f(1) = 1^2 - 1 = 0$.

STRETCHIT!

If $f(x) = x^2 = y$, can you
express x as a function
of y, $f(y)$?

Apply the function nearest
the x first and then apply the
second function to the answer
$fg(x) = f(g(x))$.

NAILIT!

$f^{-1}(x)$ is the inverse of $f(x)$.

Iterative methods

(1) Show that the equation

$$2x^3 - 2x + 1 = 0$$

has a solution between -1 and -1.5. (3 marks, ★★★★)

NAILIT!

Iterative methods can be used to solve equations that are difficult to solve using other methods. They can also be used to show that a solution to an equation lies between two values.

NAILIT!

If f(x) can take any value between a and b, then if there is a change of sign between f(a) and f(b), a root of f(x) = 0 lies between a and b.

(2) A sequence is generated using the iterative formula

 $x_{n+1} = x_n^3 + \frac{1}{9}$.

Starting with $x_0 = 0.1$, find x_1, x_2, x_3. For each of these, write down your full calculator display. (3 marks, ★★★★★)

$x_1 = $ ....

$x_2 = $ ....

$x_3 = $ ....

(3) The cubic equation $x^3 - x - 2 = 0$ has a root α between 1 and 2. (★★★★★)

The iterative formula

$$x_{n+1} = (x_n + 2)^{\frac{1}{3}}$$

with $x_0 = 1.5$, can be used to find α.

a Calculate x_4. Give your answer to 3 decimal places. (4 marks)

$x_4 = $ ....

b Prove that this value is also the value of α correct to 3 decimal places. (2 marks)

[Total: 6 marks]

(4) **a i** Sketch the graphs of $y = x^3$ and $y = 3 - x$. (2 marks, ★★★★★)

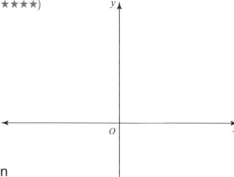

ii Hence write down the number of roots of the equation
$x^3 + x - 3 = 0$. (1 mark, ★★★★★)

....

b The cubic equation $x^3 + x - 3 = 0$ has a root α between 1 and 2.

The iterative formula

$$x_{n+1} = (3 - x_n)^{\frac{1}{3}}$$

with $x_0 = 1.2$, can be used to find α. Calculate x_6. Give your answer to 4 decimal places. (3 marks)

$x_6 = $ ....

[Total: 6 marks]

Equation of a straight line

(1) One of these graphs has the equation $y = 3 - 2x$.

State the letter of the correct graph. (1 mark, ★)

....

A 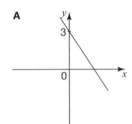   **B** 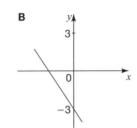   **C** 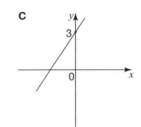   **D**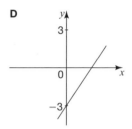

(2) **a** The line AB cuts the x-axis at 3 and the y-axis at 4.

Find the gradient of line AB. (2 marks, ★)

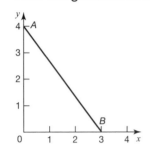

....

A straight line passes through the points $C(-3, 5)$ and $D(5, 1)$.

b Find the equation of line CD. (3 marks, ★★★★)

....

c The midpoint of *CD* is *M*. A straight line is drawn through point *M* which is perpendicular to the line *CD*.

Find the equation of this line. (3 marks, ★★★★)

NAILIT!

When two lines are perpendicular to each other, the product of their gradients is -1.

....

[Total: 8 marks]

(3) The gradient of line *OP* is 3. Its length is 12. (4 marks, ★★★★)

You will need to use Pythagoras' theorem.

Find the coordinates of point *P*.
Give each coordinate to 1 decimal place.

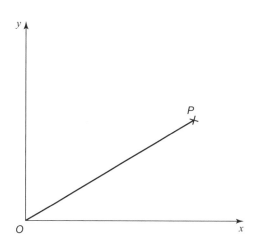

....

Quadratic graphs

(1) **a** Using the method of completing the square, solve this quadratic equation.

Give your answers to 1 decimal place.

$x^2 + 4x + 1 = 0$ (4 marks, ★★★★★)

SNAPIT! Quadratic graphs

Quadratic graphs are ∪-shaped if the coefficient of x^2 is positive and ∩-shaped if the coefficient is negative.

b Hence sketch the graph of $y = x^2 + 4x + 1$ on the axes provided, including the coordinates of the turning point. (3 marks, ★★★★★)

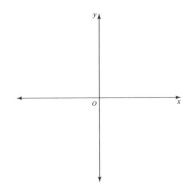

NAILIT!

To find the turning point of a quadratic graph, first complete the square to get the equation in the form $y = a(x + p)^2 + q$.

The turning point is at $(-p, q)$.

[Total: 7 marks]

(2) Express $5x^2 - 20x + 10$ in the form $a(x + b)^2 + c$.

Write down the values of a, b and c. (4 marks, ★★★★)

You will need to take 5 out as a factor as part of the process of completing the square.

(3) Express $2x^2 + 12x + 3$ in the form $a(x + b)^2 + c$.

Write down the values of a, b and c. (4 marks, ★★★★)

Recognising and sketching graphs of functions

(1) Fill in the table by inserting the letter of the graph that fits with the equation. (6 marks, ★★★★)

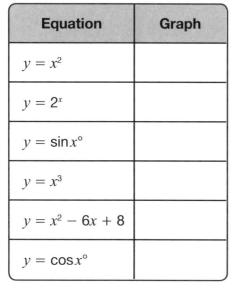
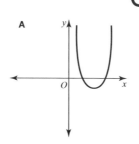
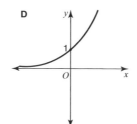
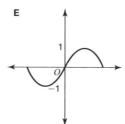
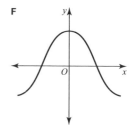
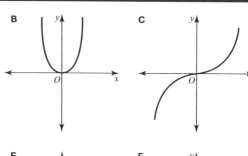

Equation	Graph
$y = x^2$	
$y = 2^x$	
$y = \sin x°$	
$y = x^3$	
$y = x^2 - 6x + 8$	
$y = \cos x°$	

(2) **a** On the axes provided, sketch the graph of $y = \sin x$ for $0° \leq x \leq 360°$. (3 marks, ★★★★)

There is no scale on the y-axis. You need to mark the maximum and minimum values on the axis.

b On the axes provided, sketch the graph of $y = \tan x$ for $0° \leq x \leq 360°$. (3 marks, ★★★★)

[Total: 6 marks]

(3) Find all the values of θ in the range $0° \leq \theta \leq 360°$ that satisfy

$3\cos\theta = 1$.

Give your answers to 1 decimal place. (3 marks, ★★★★★)

Translations and reflections of functions

NAILIT!

$y = f(x) \rightarrow f(x + a)$:
translation of $-a$ units
parallel to the x-axis

$y = f(x) \rightarrow f(x) + a$:
translation of a units
parallel to the y-axis

$y = f(x) \rightarrow -f(x)$:
reflection in the x-axis

$y = f(x) \rightarrow f(-x)$:
reflection in the y-axis

(1) The quadratic curve $y = f(x)$ passes through the origin and the point (4, 0), and has a turning point at (2, −4). On the same axes, sketch the graphs of (★★★★)

a $y = -f(x)$ (2 marks)

b $y = f(x - 2)$ (2 marks)

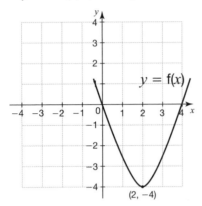

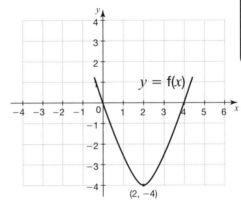

In each case, give the coordinates of the points of intersection of the graph with the x-axis and the coordinates of the turning point.

NAILIT!

A turning point is a maximum or minimum point on the curve.

....   **[Total: 4 marks]**

(2) The graph of $y = f(x)$ is shown on the right.

On the same axes, sketch the graphs of these functions. (★★★★)

a $y = -f(x)$ (2 marks)   **c** $y = f(-x)$ (2 marks)

b $y = f(x) + 2$ (2 marks)

[Total: 6 marks]

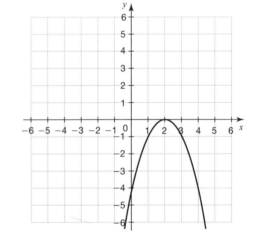

(3) On the axes below, sketch the graph of $y = \cos(x) + 2$ for $0° \leq x \leq 360°$. (2 marks, ★★★★★)

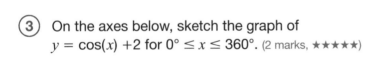

Equation of a circle and tangent to a circle

NAILIT!

A circle with centre the origin (0, 0) and radius r has the equation $x^2 + y^2 = r^2$.

(1) Write down the radius of each of these circles. (★★★)

a $x^2 + y^2 = 25$ (1 mark)   **b** $x^2 + y^2 - 49 = 0$ (1 mark)   **c** $4x^2 + 4y^2 = 16$ (2 marks)

....

[Total: 4 marks]

(2) A circle has the equation $x^2 + y^2 = 21$. Determine whether the point (4, 3) lies inside or outside this circle. (4 marks, ★★★★★)

> Compare the length of line joining the origin to the point (4, 3) with the radius of the circle.

....

(3) The point $P(5, 7)$ lies on a circle with centre the origin. (★★★★★)

a Find the radius of the circle.
Give your answer as a surd. (2 marks)

WORKIT!

The point $P(2, 3)$ lies on a circle with centre the origin, O.

Find the equation of the tangent to the circle at the point.

1 Find the gradient of the radius OP. $\frac{3}{2}$

2 Find the gradient of the tangent at P. $-\frac{2}{3}$

3 Use $y - y_1 = m(x - x_1)$.

$y - 3 = -\frac{2}{3}(x - 2)$ so
$y = -\frac{2}{3}x + \frac{13}{3}$

....

b Write down the equation of the circle. (1 mark)

....

c Find the equation of the tangent to the circle at point P. (3 marks)

NAILIT!

A tangent is a line which just touches the circle.

....

[Total: 6 marks]

Real-life graphs

(1) The velocity–time graph shows the motion of an object. (★★★)

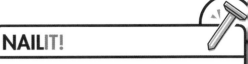

NAILIT!

In a distance–time graph,
gradient = velocity (or speed).

In a velocity–time (or speed–time) graph,
gradient = acceleration
area under graph = distance travelled.

a Calculate the acceleration of the object during the first 10 seconds.
(2 marks)

....

b Calculate the total distance travelled by the object. (2 marks)

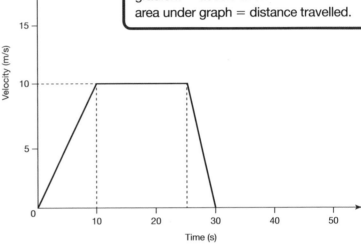

....

[Total: 4 marks]

(2) This is the speed–time graph for a car journey. (★★★★)

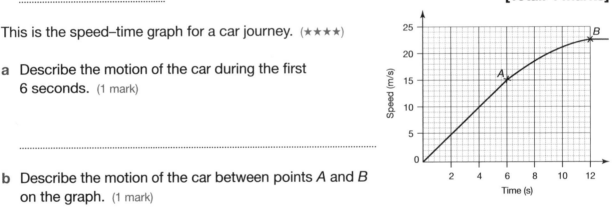

a Describe the motion of the car during the first 6 seconds. (1 mark)

....

b Describe the motion of the car between points *A* and *B* on the graph. (1 mark)

....

....

c Work out an estimate for the distance travelled by the car between points *A* and *B* on the graph by using three trapeziums. Give your answer to the nearest integer. (3 marks)

NAILIT!

To estimate the area under a curve, divide the area into trapeziums.

....

d Giving a reason, say whether your answer for the distance is an underestimate or an overestimate of the actual distance. (2 marks)

....

....

[Total: 7 marks]

Generating sequences

NAILIT!

You need to be familiar with arithmetic and geometric sequences, as well as the sequences of square numbers, cube numbers, triangular numbers and Fibonacci numbers.

(1) **a** Work out the next term in each of these sequences. (★★)

    i 16, 8, 4, 2, 1, ... (1 mark)   **iii** 5, 9, 13, 17, ... (1 mark)

....

    ii 3, 9, 27, 81, ... (1 mark)

Look for adding or subtracting the same number to get from one term to the next. Then look for multiplying or dividing by the same number.

....

  b The following sequence is an arithmetic sequence.

    Work out the missing two terms. (★★)

    $27, ..., ..., -12$ (2 marks)

Work out the difference between 27 and -12. Then divide your answer by three, because three equal distances give two new terms (in the gaps between them).

....   **[Total: 5 marks]**

NAILIT!

A term-to-term rule tells you how to work out the next term in the sequence from the current term.

(2) The first two terms of a sequence are

3, 1, ..., ...

The term-to-term rule for this sequence is to multiply the previous term by 2 and subtract 5.

Work out the next two terms of the sequence. (2 marks, ★★)

First work out the difference between the two terms you are given.

.... , ...

(3) Write down the next two terms for these sequences. (★★★)

  a 1, 4, 9, 16, ..., ... (1 mark)

.... , ...

  b 1, 3, 6, 10, ..., ... (1 mark)

.... , ...

  c 1, 1, 2, 3, 5, ..., ... (1 mark)

NAILIT!

Not all sequences have a constant difference or multiplier. Try squares, cubes, triangular numbers and Fibonacci sequences.

.... , ...

[Total: 3 marks]

The nth term

1 The first four terms of an arithmetic sequence are

2, 6, 10, 14, ... (★★★)

> **NAILIT!**
>
> You need to be able to write the nth term of linear and quadratic sequences as an expression containing n.

a Find an expression for the nth term of this sequence. (2 marks)

....

b Use your answer to part a to explain why all the terms of this sequence are even. (2 marks)

....

....

c Work out whether 236 is a number in this sequence. (2 marks) ◄

> Write 236 equal to the nth term.

....

[Total: 6 marks]

2 An expression for the nth term of a sequence is $9 - n^2$. (★★★)

a Find the 2nd term of this sequence. (1 mark)

....

b Find the 20th term of this sequence. (1 mark)

....

c Explain why 10 cannot be a term in this sequence. (1 mark)

....

....

[Total: 3 marks]

3 Find the nth term for the sequence

1, 1, 3, 7, 13, ... ◄

(5 marks, ★★★★★)

> There is no constant difference here, so it is likely that this is a quadratic sequence. You need to find the 1st and 2nd differences.

....

> **WORKIT!**
>
> Find the nth term of the sequence   2   3   8   17 ...
>
> **1** Work out the 1st differences.   1   5   9
>
> **2** Work out the 2nd differences.   4   4
>
> **3** The coefficient of n^2 is $\frac{1}{2} \times$ 2nd difference.
>
> nth term starts $2n^2$
>
> **4** Subtract $2n^2$ from each term.   0   −5   −10   −15
>
> **5** Work out the nth term for sequence in step 4. $-5n + 5$
>
> **6** Combine the two expressions from steps 4 and 5.
>
> nth term is $2n^2 - 5n + 5$.

Arguments and proofs

(1) Sarah says that all prime numbers are odd. By giving a counter-example, show that her statement is false. (1 mark, ★★)

....

NAILIT!

Proof questions can be on many topics but the majority of them can be solved algebraically. Often you need to substitute a letter to prove an expression. Sometimes you can prove something is not true by giving a counter-example (i.e. a situation when it is false).

(2) Explain whether each of these statements is true or false. (★★★)

    a If n is a positive integer, $2n + 1$ is always greater than or equal to 3. (1 mark)

....

....

    b $3(n + 1)$ is always a multiple of 3. (1 mark)

....

....

    c If n is a positive integer, $2n - 3$ is always even. (1 mark)

....

....

[Total: 3 marks]

(3) Prove that the sum of any two consecutive integers is always an odd number. (3 marks, ★★★★)

> Start by letting the first number be x. Then write an expression in terms of x for the second number.

(4) Prove that $(2x - 1)^2 - (x - 2)^2$ is a multiple of 3 for all integer values of x. (3 marks, ★★★★★)

> Multiply out the brackets and then simplify. Look for a factor of 3.

(5) Prove that the difference between the squares of two consecutive odd numbers is always a multiple of 8. (5 marks, ★★★★★)

Ratio, proportion and rates of change
Introduction to ratios

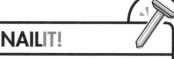

1 A bag contains red balls and black balls only. The number of black balls to the number of red balls is in the ratio 3:2.

There are 18 black balls. Work out the number of balls in the bag. (3 marks, ★) ◄──── *You know 18 black balls are equal to 3 parts.*

NAILIT!

In ratio questions, find the total number of parts by adding the numbers in the ratio together.

....

2 Three daughters are aged 15, 17 and 18 years. They are left £25 000 in a will to be divided between them in the ratio of their ages. Calculate how much each daughter will receive. (3 marks, ★)

WORKIT!

Divide 3.5 kg in the ratio 4:3

1 Work out the total numbers of parts: $3 + 4 = 7$.

2 Work out the value of one part: $3.5 \div 7 = 0.5$ kg.

3 Work out the value of each part of the ratio: 4×0.5 and 3×0.5.

2 kg and 1.5 kg

....

3 A farm has total area of 800 acres. 40% of the area is devoted to arable crops. The rest is devoted to cattle and sheep in the ratio 9:7. ◄──── *First work out the area devoted to livestock (60% of 800 acres). Divide your answer in the ratio given.*

Work out the land area in acres devoted to sheep. (3 marks, ★★★)

....

4 Given that $3x + 1:x + 4 = 2:3$, find the value of x. (3 marks, ★★★★★)

....

5 A wood has pine, oak and ash trees.

The numbers of pine trees and oak trees are in the ratio 5:8.

The numbers of oak trees and ash trees are in the ratio 2:3. ◄──── *Using a multiplier for the second ratio, find a new three-part ratio for all three tree types. Then work out how many parts there are in total.*

The total number of trees in the wood is 300. Find the number of ash trees. (4 marks, ★★★★★)

....

Scale diagrams and maps

(1) The scale on a map is 1:50000.

Two towns are 10cm apart on the map.

What is their actual distance apart?
Give your answer in km. (2 marks, ★)

....

NAILIT!

When changing a map scale to a ratio, make sure that the two quantities in the ratio are changed into the same units. For example,
1cm:500m = 1cm:50000cm = 1:50000.

(2) A map is drawn to a scale of 1:40000. Find the actual length, in km, of a straight road of length (★)

a 2.3cm (1 mark)   **b** 3mm (1 mark) ◄──

Multiply each of these lengths by 40000 and then change the units.

....

[Total: 2 marks]

(3) On a scale drawing of a garden, the length of a path whose actual distance is 40m is 5cm. Find the scale of the drawing in the form 1:n where n is an integer. (3 marks, ★★)

....

STRETCHIT!

If you photocopied this page at A3 scale, the enlargement could be expressed by the ratio 1:$\sqrt{2}$.

How would this affect the answer you get in Question 4?

(4) The map shows a port and a gas rig.

The actual distance from the port to the gas rig is 12km.

By taking a measurement from the map, work out the scale of the map.

Give your answer in the form 1:n where n is an integer. (3 marks, ★★)

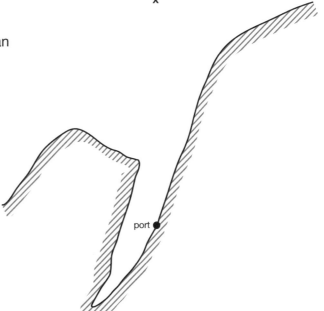

....

Percentage problems

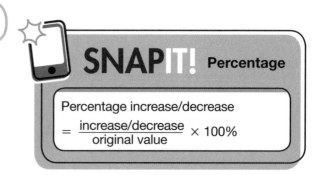

SNAPIT! Percentage

Percentage increase/decrease
$$= \frac{\text{increase/decrease}}{\text{original value}} \times 100\%$$

(1) A store selling bikes increases the price of a bike from £350 to £385. Find the percentage increase. (2 marks, ★★)

....

(2) A football manager initially earns £600 000 per year. On the promotion of his team to the Premier League his earnings increase to £1.1 million per year. Find the percentage increase in his pay to 1 decimal place. (2 marks, ★★)

....

STRETCHIT!

The Whizzie skateboard has been reduced by 20%, from £22 to £17.60. Karl bought his Zoomtown board somewhere else, also for £17.60. He tells Katie, 'At full price, Whizzie skateboards are 20% more expensive than Zoomtown ones'. Is he right?

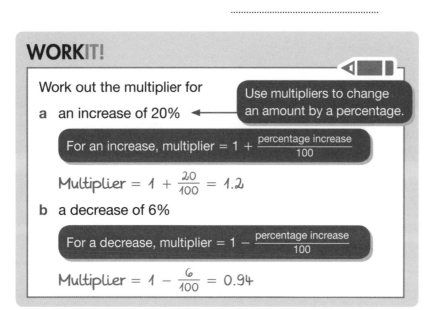

WORKIT!

Work out the multiplier for

> Use multipliers to change an amount by a percentage.

a an increase of 20%

> For an increase, multiplier = $1 + \frac{\text{percentage increase}}{100}$

Multiplier = $1 + \frac{20}{100} = 1.2$

b a decrease of 6%

> For a decrease, multiplier = $1 - \frac{\text{percentage increase}}{100}$

Multiplier = $1 - \frac{6}{100} = 0.94$

(3) After 3 years a caravan originally costing £25 000 has decreased in value by 28%. What is its new value? (2 marks, ★★)

....

(4) The price of a motorbike after a reduction of 12% is £14 300. Find the original price of the motorbike. (3 marks, ★★★)

> Start by saying that 88% of the original price is £14 300.

....

(5) Fran invests £8000 at an interest rate of 2.8% for 4 years. If simple interest is paid, find the total amount of interest paid over the 4 years. (2 marks, ★★★)

....

Direct and inverse proportion

(1) The pressure of a gas, P, is directly proportional to its temperature, T. (★★★)

a Write the above statement as an equation. (1 mark)

NAILIT!

When y is directly proportional to x you can write $y = kx$.

When y is inversely proportional to x you can write $y = \frac{k}{x}$

....

b The pressure is 200 000 Pascals when the temperature is 540 Kelvin.

Find the value of pressure when the temperature is 200 Kelvin.
Give your answer to the nearest whole number. (3 marks)

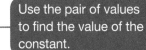

Use the pair of values to find the value of the constant.

....

[Total: 4 marks]

(2) The cost of building a circular garden pond is directly proportional to the square of the radius of the pond. The cost when the radius is 3 m works out at £480.

Find the cost of building a circular pond with a radius of 4 m. Give your answer to the nearest whole number. (4 marks, ★★★★)

....

(3) A quantity c is inversely proportional to another quantity h.

When $c = 3$, $h = 12$ (★★★)

a write a formula for c in terms of h.
(1 mark)

b calculate the value of c when $h = 15$.
(2 marks)

....

[Total: 3 marks]

(4) Emma goes to France on holiday. (★★★)

a She changes £350 into euros at an exchange rate of £1 = €1.15

Work out how many euros she gets. (1 mark)

....

b When she returns she still has €80.

She changes this back to pounds at an exchange rate of £1 = €1.11.

How many pounds does she get? Give your answer to the nearest penny. (2 marks)

....

c How much would she have saved if she had only changed the money that she needed for the holiday? (3 marks)

....

[Total: 6 marks]

Graphs of direct and inverse proportion and rates of change

① Which one of these graphs shows that y is directly proportional to x? (1 mark, ★★★)

A 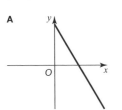 B 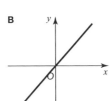 C 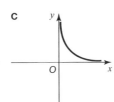 D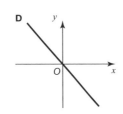

....

NAILIT!

Make sure you can recognise and draw graphs showing direct and inverse proportion.

The graph showing direct proportion is a straight line passing through the origin with a positive gradient. The graph showing inverse proportion is a curve that does not cross either axis.

② The pressure, P, of a gas is inversely proportional to the volume, V. Which graph correctly shows this statement? (1 mark, ★★★)

A 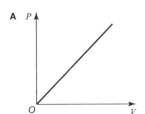 C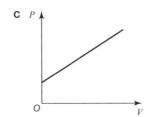

B 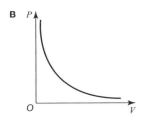 D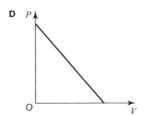

....

③ In a chemical experiment, the mass of contents in a flask was measured every minute and the loss in mass calculated. The results are shown in the table. (★★★★)

Time (minutes)	0	1	2	3	4	5	6	7
Loss in mass (g)	0	9.8	19.6	24.5	26.6	27.7	28	28

a Plot the results on the grid. (2 marks)

b Work out the initial rate of loss of mass in

  i g/minute (2 marks)

  .... g/minute

  ii g/second. (1 mark)

  .... g/second

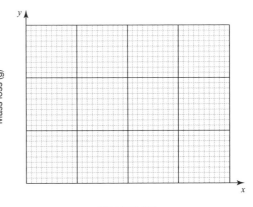

Mass loss (g)

Time (minutes)

[Total: 5 marks]

Growth and decay

NAILIT!

If a quantity grows by a set percentage each time interval (e.g. hour, year),

multiplier $= 1 + \frac{\% \text{ increase for each time unit}}{100}$

If a quantity decays/falls by a set percentage each time interval (e.g. hour, year),

multiplier $= 1 - \frac{\% \text{ decrease for each time unit}}{100}$

Amount at the end of n time units $= A_0 \times (\text{multiplier})^n$, where $A_0 =$ original amount

---

(1) The population of a town is 150 000. Each year the population increases by 6%. (★★★★)

> As the population is increasing, the multiplier will be greater than 1.

   a What will the population be 3 years from now? (2 marks)

     ....

   b After how many years will the population have risen to over 200 000? Give your answer to the nearest year. (2 marks)

     ....

[Total: 4 marks]

(2) Jenny buys an electric car. The car costs £21 000 and it depreciates at a rate of 12% each year. What will be the car's value after 4 years? Give your answer to the nearest whole number. (3 marks, ★★★★)

> Depreciation means that the multiplier will be less than 1.

     ....

(3) A radioactive isotope halves its activity every 12 seconds.

The initial activity of a sample of the isotope was 100 units.

Find the activity after 2 minutes. Give your answer to 1 significant figure. (4 marks, ★★★★★)

     ....

Ratios of lengths, areas and volumes

NAILIT!

When using scale factors to work out lengths, areas or volumes, you must make sure that the two shapes are similar.

NAILIT!

If two shapes A and B are similar, we can say l_A is one of the lengths on shape A and l_B is the corresponding length on shape B.

The scale factor for lengths is $\frac{l_B}{l_A}$.

The scale factor for the area A from A to B can be written:

$$\frac{A_B}{A_A} = \left(\frac{l_B}{l_A}\right)^2$$

The scale factor for the volume V from A to B can be written:

$$\frac{V_B}{V_A} = \left(\frac{l_B}{l_A}\right)^3$$

(1) These two triangular prisms are mathematically similar. (★★★★)

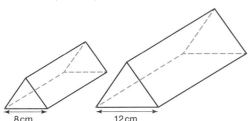

8 cm   12 cm

a Find the scale factor for the volume of the larger prism in relation to the smaller one. (2 marks)

Work out the scale factor for the lengths first, then use this to find the scale factor for volume.

....

b The area of the triangular cross-section of the small prism is 10 cm². Find the area of cross-section of the large prism. (2 marks)

.... cm²

c The volume of the large prism is 450 cm³. Find the volume of the small prism. Give your answer to the nearest whole number. (2 marks)

.... cm³

WORKIT!

Cylinder A and cylinder B are mathematically similar.

The radius of cylinder B is 1.5 times the radius of cylinder A.

The volume of cylinder A is 14 m². Work out the volume of cylinder B.

Scale factor for length = 1.5
Scale factor for volume = 1.5³
Volume of cylinder B = 14 × 1.5³
= 47.25 m³

[Total: 6 marks]

(2) These two solid cuboids are mathematically similar.

The volume of the larger cuboid is 195.3% of the volume of the smaller cuboid.

Calculate the height, h, of the larger cuboid.
Give your answer to the nearest cm. (4 marks, ★★★★)

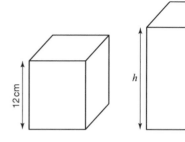

....

(3) *XY*, *TU* and *WZ* are parallel lines. *YZ*, *YW* and *XZ* are straight lines.

$YU = 10$ cm, $UZ = 5$ cm and $UT = 3$ cm. (★★★★★)

a Calculate the length of

    i *XY* (1 mark)   ii *WZ* (1 mark)

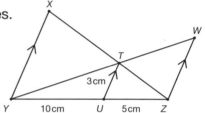

.... cm   .... cm

b Find the ratio of the areas of triangles *TYX* and *TWZ*.
Give your answer in the form $n:1$ (2 marks)

> If the area of triangle *TWZ* is 1 unit, n represents the comparable area of *TYX*.

....

[Total: 4 marks]

Gradient of a curve and rate of change

1 The velocity–time graph shows a car's motion for 90 seconds. (★★★★★)

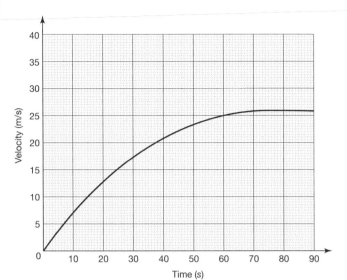

a Find the acceleration during the first 15 seconds. (2 marks)

....

b Work out the instantaneous acceleration at 45 seconds. (2 marks)

....

c Work out the average acceleration over the first 70 seconds. (2 marks)

....

d At a certain time the instantaneous acceleration is equal to the average acceleration over the first 70 seconds. (2 marks)

Find the time in seconds when this happens.

....

[Total: 8 marks]

Converting units of areas and volumes, and compound units

① The formula to work out pressure is pressure $= \dfrac{\text{force}}{\text{area}}$.

Calculate the pressure produced by a force of 200 N acting on an area of 0.4 m². Give your answer in N/m². (1 mark, ★★)

The answer requires the force to be in Newtons and the area in m² so no conversions are needed.

....

② Calculate the pressure in Newton/m² if a force of 500 Newtons acts on an area of 200 cm². (2 marks, ★★★)

....

③ The formula for density is density $= \dfrac{\text{mass}}{\text{volume}}$.

Copper has a density of 8.92 g/cm³.

Find the mass in grams of a length of copper wire with a volume of 12 cm³.
Give your answer to the nearest gram. (2 marks, ★★★)

> **NAILIT!**
>
> Compound units are made up of two units: for example m/s for speed, km/h for speed, g/cm³ for density. When entering units into a formula, check whether the values are in the units needed for the final answer. If not, you will need to convert your answer to the correct units.

....

④ A picture has a length of 167 cm and a width of 54 cm.

Joshua wants to work out the area of the picture in m² correct to 2 decimal places.

Here is his working.

$$\text{Area} = \text{length} \times \text{width} = 167 \times 54 = 9018 \, \text{cm}^2$$
$$\text{Area in m}^2 = \frac{9018}{100} = 90.18 \, \text{m}^2$$

Joshua's answer is wrong.

Explain what he has done wrong and work out the correct area in m². (2 marks, ★★★)

....

⑤ Umar drove for 2 hours at 60 km/h. He then drove for 3 hours at 80 km/h.

Work out his average speed for the journey. (3 marks, ★★★★)

....

Geometry and measures
2D shapes

(1) State whether each of these statements is true or false. (★)

a The diagonals of a rhombus cut each other at right-angles. (1 mark)

....

b The diagonals of a parallelogram are always the same length. (1 mark)

....

c A kite has one line of symmetry. (1 mark)

....

d A cuboid has 8 vertices and 6 faces. (1 mark)

....

e A trapezium has one pair of parallel sides. (1 mark)

....

f A pentagon has 5 sides of equal length. (1 mark)

.... **[Total: 6 marks]**

(2) Identify these shapes (★★)

a four equal sides, two pairs of equal angles (1 mark)

....

b two pairs of equal sides, different length diagonals (1 mark)

....

c three sides, order of rotational symmetry 3 (1 mark)

....

d four sides, one line of symmetry, 1 pair of equal angles (1 mark)

.... **[Total: 4 marks]**

Constructions and loci

(1) Draw the locus of the point that is always the same distance from points *A* and *B*. (2 marks, ★★)

> Construct the perpendicular bisector of *AB*. ⟶   *A* ———————————— *B*

(2) A goat is secured by a chain of length 1.5 m and a loop on a bar of length 6 m. The loop can travel anywhere along the length of the bar.

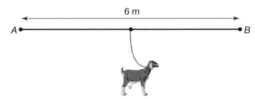

Draw an accurate diagram using a scale of 1 cm to 1 m to show the area in which the goat can graze. Shade the area. (3 marks, ★★★)

(3) The diagram below shows two walls, *XY* and *XZ*, at an angle to each other.

A spider waits at point *X*. It then moves along a path such that its distance to lines *XY* and *YZ* is the same. (★★★)

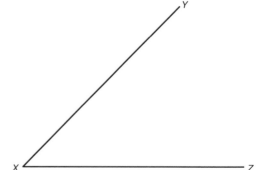

a Using only compasses and a ruler, draw the path of the spider on the above diagram. You should show all your construction lines. (2 marks)

b A fly walks along a path between the two walls so that the distance from point *X* is always 6 cm. Mark the point between the walls where it is possible for the spider's and the fly's paths to cross. (2 marks)

[Total: 4 marks]

Properties of angles

Sometimes shapes in exam questions may be deliberately drawn slightly inaccurately ('not drawn to scale'). For example, they may look like a regular shape when they are not. The diagram below looks like a rhombus, but you would have to prove it using the information in the question and diagram.

NAILIT!

Look for:

- alternate or corresponding angles in parallel lines
- angles in a triangle, including isosceles and equilateral triangles
- angles on a straight line.

(1) *ABCD* is a rhombus. Point *O* is the point where the diagonals intersect. (★★)

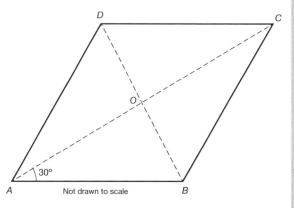

a State, giving a reason, the size of angle *ACB*. (1 mark)

....

....

b State, giving a reason, the size of angle *AOB*. (1 mark)

....

....

Think about the properties of a rhombus.

c Work out the size of angle *BDC*. You must show your workings. (2 marks)

.... °

[Total: 4 marks]

(2) *ABC* is an isosceles triangle with *AC = BC* and angle *ACB* = 36°. *BD* is perpendicular to *AC*.

Find the size of angle *ABD*. Give reasons for each stage of your working. (2 marks, ★★)

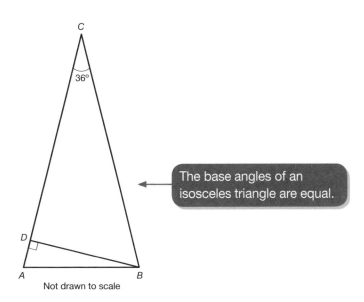

The base angles of an isosceles triangle are equal.

.... °

(3) *AB*, *CD* and *XY* are straight lines. (★★★)

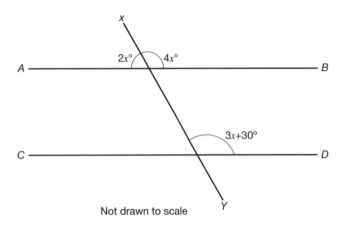

Not drawn to scale

a Find the value of x. You must show all your workings. (2 marks)

.... °

b Prove that lines *AB* and *CD* are parallel. (2 marks)

[Total: 4 marks]

(4) A needlework group is designing a patchwork quilt. The diagram shows part of their design – a square aligned with a regular pentagon.

Calculate the size of angle x. (4 marks, ★★★)

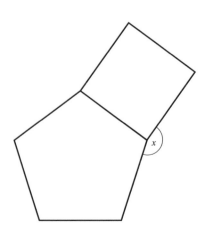

.... °

Congruent triangles

(1) *ABCD* is a parallelogram.

Prove that angle *BAD* = angle *BCD*. (3 marks, ★★★)

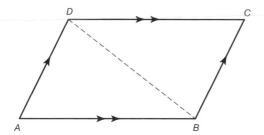

SNAP IT!

You can show that triangles are congruent by demonstrating identical values for:

• three sides (SSS)

• two sides and the angle between them (SAS)

• two angles and one side (ASA)

• right angle, hypotenuse and one other side (RHS).

Start by listing the sides and angles that are the same for both triangles.

(2) *ABC* is an isosceles triangle. *AB* = *AC*. Prove that the perpendicular from *A* to *BC* bisects *BC*. (3 marks, ★★★) ← If you are not given a diagram, sketch one.

(3) *ABC* is a right-angled triangle. *P* is the midpoint of *AB* and *Q* is the midpoint of *BC*. *OP* is parallel to *BC* and *OQ* is parallel to *AB*.

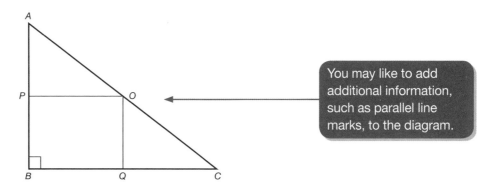

You may like to add additional information, such as parallel line marks, to the diagram.

Prove that triangles *AOP* and *OCQ* are congruent. (4 marks, ★★★★★)

Transformations

1. Describe the single transformation that maps triangle A onto triangle B. (1 mark, ★)

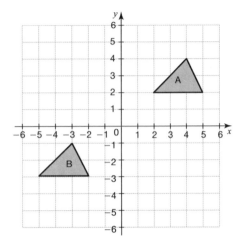

2. On the grid below, enlarge the triangle by a scale factor of −3 using centre (−3, 3). (3 marks, ★★★★)

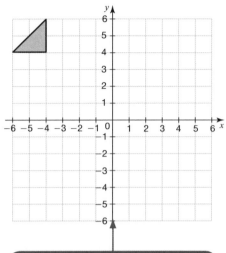

A negative scale factor means that the image is on the other side of the centre of enlargement.

....

3. Triangle A is shown on the following grid. (★)

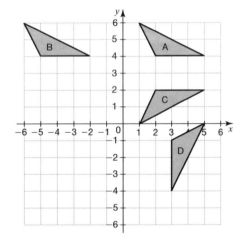

NAILIT!

You need to know how to draw and describe translations, rotations, reflections and enlargements.

a Describe the single transformation that maps triangle A onto triangle B. (1 mark)

....

b Describe the single transformation that maps triangle A onto triangle C. (1 mark)

....

c Describe the single transformation that maps triangle A onto triangle D. (2 marks)

....

[Total: 4 marks]

Invariance and combined transformations

(1) **a** Triangle A is reflected in the line $x = 0$.
How many invariant points are there on the triangle? (1 mark, ★★★)

....

b **i** Triangle A is rotated 90° anticlockwise about the origin. The resulting triangle is then translated by $\binom{6}{0}$.

Draw this new triangle and label it B. There is one invariant point on triangle B. Give the coordinates of this point.

(2 marks, ★★★)

....

ii Describe the single transformation that maps triangle A to triangle B. (1 mark, ★★★)

....

[Total: 4 marks]

(2) **a** Triangle *PQR* is reflected in the line $x = 4$ and then rotated 90° anticlockwise about point *R*. Draw the image after these two transformations on the grid. (2 marks, ★★★★)

Use tracing paper to work out the position of a shape after a rotation.

b Give the coordinates of the invariant point after the two transformations.
(1 mark, ★★★★)

....

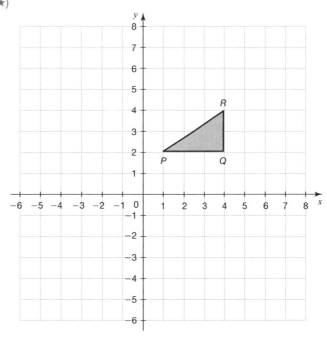

[Total: 3 marks]

3D shapes

(1) Some 3D shapes are drawn below.

A

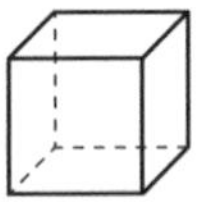

B

C

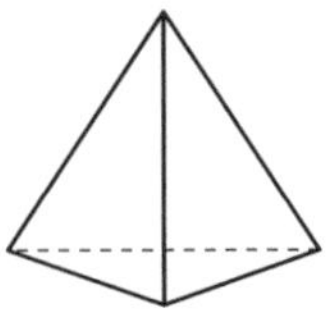

D

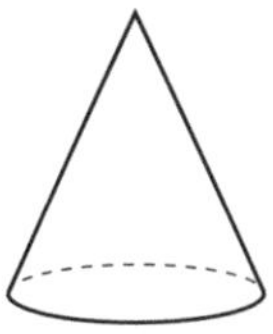

E

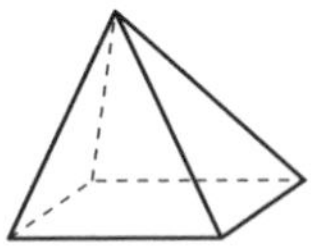

F

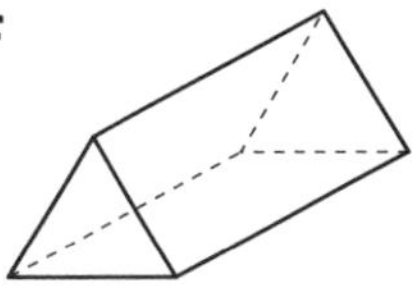

G

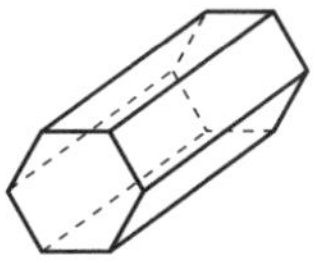

H

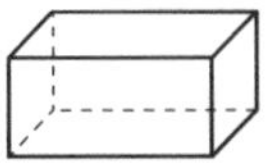

NAILIT!

You need to recognise and be able to sketch these 3D shapes: cube, cuboid, triangular prism, hexagonal prism, cylinder, square-based pyramid, triangular-based pyramid, cone and sphere.

There are unlikely to be questions based only on 3D shapes, but they will crop up in other topics.

STRETCHIT!

If you had a shape made up of two pentagons on either end, joined by five rectangles, what would it be called? How many vertices (corners) and edges would it have?

Give the letter or letters of the shapes that fit these descriptions. (★)

a The shape with 8 faces. (1 mark)

....

b The shapes with no vertices. (1 mark)

....

c The two shapes with 8 vertices. (1 mark)

....

d The shape with no faces, no edges and no vertices. (1 mark)

....

e The shape with 6 edges. (1 mark)

....

f The two shapes with 12 edges. (1 mark)

....

[Total: 6 marks]

Parts of a circle

(1) The diagram shows a circle with centre O.

Name the lines labelled a to d. (★)

a .... (1 mark)

b .... (1 mark)

c .... (1 mark)

d .... (1 mark)

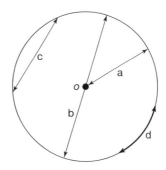

[Total: 4 marks]

(2) Here are some circles with centre O.

For each circle, give the correct name for the shaded area. (★★)

a (1 mark)

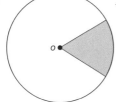

Remember to use the words **major** and **minor** when describing sectors and segments.

....

b (1 mark)

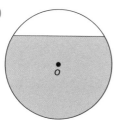

....

c (1 mark)

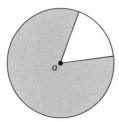

....

d (1 mark)

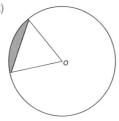

....

[Total: 4 marks]

Circle theorems

WORKIT!

The line *DCE* is a tangent to the circle at *C*. *B*, *C* and *F* lie on a straight line. *AB* and *DE* are parallel. Angle *DCF* = 40°.

Work out the size of angle *ACB*.

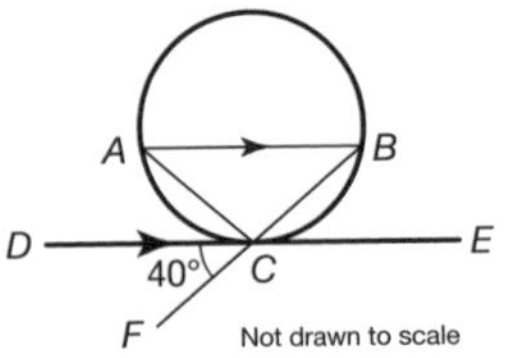

Not drawn to scale

Angle *ABC* = angle *DCF* = 40° (corresponding angles)
Angle *BCE* = angle *DCF* = 40° (vertically opposite angles)
Angle *BAC* = angle *BCE* = 40° (alternate segment theorem)
Angle *ACB* = 180 − (40 + 40) = 100° (angle sum in a triangle)

NAILIT!

Make sure that you have learnt all the circle theorems:

- angles in the same segment
- angle in a semicircle
- angles in a cyclic quadrilateral
- alternate segment theorem.

1 The line *TB* is a tangent to the circle at *T*. The line *AB* is a straight line passing through the centre of the circle, *O*. Angle *OBT* = 28°.

The angle between a tangent and the radius is 90°.

Work out the size of angle *OAT*. Give a reason for each stage of your working. (4 marks, ★★★)

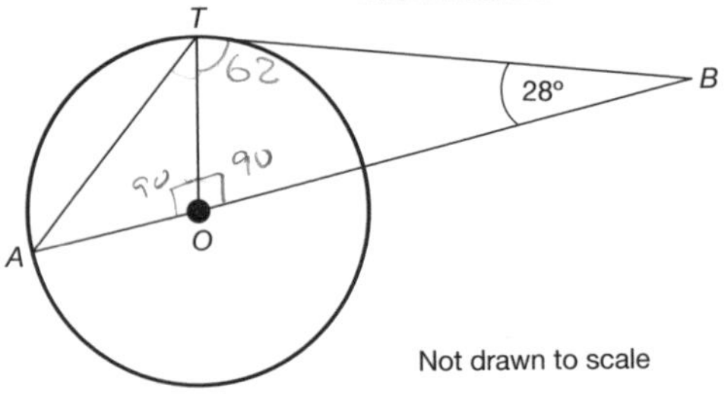

Not drawn to scale

90 + 28 = 118
180 − 118 = 62
$\frac{62}{2}$ = 31°

31 °

2 *A*, *B* and *C* are points on the circumference of a circle with centre *O*. A tangent to the circle is drawn at point *B*. (★★★★)

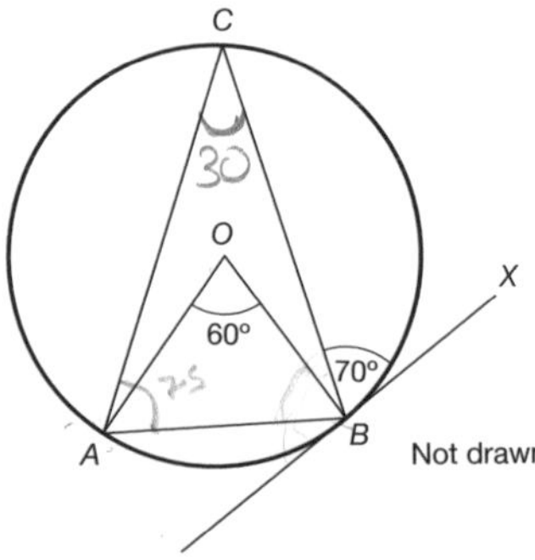

Not drawn to scale

a Write down the size of angle *ACB*. Give a reason for your answer. (2 marks)

30 because the angle at the centre is twice the angle at the circumference

b Calculate the size of angle *BAC*. Give a reason for your answer. (2 marks) 180 − 30 = 150 $\frac{150}{2}$ = 75

Learn to recognise when you can use the alternate segment theorem.

70✗ 75° because angles in a triangle add up to 180

c Calculate the size of angle *CAO*. Give a reason for your answer. (2 marks)

180 − 60 = 120 $\frac{120}{2}$ = 60 75 − 60 = 15

10✗ 15°

[Total: 6 marks]

Projections

(1) This solid shape is made out of seven one-centimetre cubes.

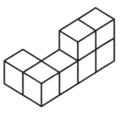

On the one-centimetre grid shown below, draw a plan view of this solid. (2 marks, ★★)

NAILIT!

The question will usually give arrows on the diagram to show the viewing direction. If you are asked for a plan view, this is the view from above.

You need to draw in all the edges where there is a change of level.

(2) This solid shape is made out of one-centimetre cubes. (★★)

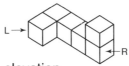

a On the grids below draw the side elevation from L and the side elevation from R. (2 marks)

b Draw the plan view of the solid on the grid below. (1 mark)

L   R

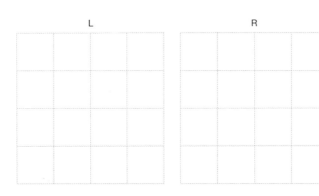

[Total: 3 marks]

(3) The diagram shows the plan, front elevation and side elevation of a solid shape on a centimetre grid.

Draw a sketch of the solid shape. Write the dimensions on your sketch. (3 marks, ★★★)

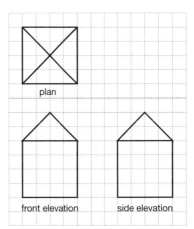

plan

front elevation   side elevation

Bearings

WORKIT!

The bearing of B from A is 285°.
What is the bearing of A from B?

1 Draw a sketch with the north arrows through both A and B.

2 Work out the angle between the north arrow and line AB at A.

3 Use alternate angles to work out the angle between the north arrow and line AB at B.

4 Use this angle to work out the bearing.

N
105°
B
N
285 °
A
285 − 180 = 105°

105°

SNAPIT!

Bearings are always measured from due north in a clockwise direction.

(1) Jack walks across the moor on a bearing of 050°. What bearing must he take in order to walk in a straight line back to his starting point? (2 marks, ★★★)

NAILIT!

If you aren't given a diagram, draw a sketch.

.... °

NAILIT!

Always draw a north line through each point. Use the angle properties of parallel lines when working out bearings.

(2) The map shows a port and a lighthouse.

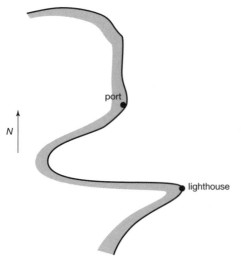

port

N

lighthouse

An offshore wind turbine is on a bearing of 100° from the port and on a bearing of 040° from the lighthouse. Mark this information on the diagram, using construction lines to show the position of the wind turbine. Mark this with an X. (3 marks, ★★★)

Pythagoras' theorem

1) The diagram shows a garden.

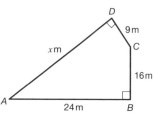

$AB = 24$ m, $BC = 16$ m and $CD = 9$ m.
Angle ABC = angle $ADC = 90°$.

Work out the perimeter of the garden. Give
your answer to the nearest metre. (4 marks, ★★★★★)

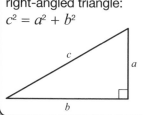

SNAP IT!

Pythagoras' theorem
states that for any
right-angled triangle:
$c^2 = a^2 + b^2$

.... m

2) $ABCDEFGH$ is a cuboid.

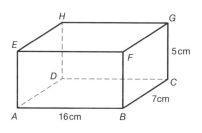

A spider starts a web by joining A to G with a straight strand of
web. It walks along AB, BC and CG to reach G.

How much further does it walk than the length of web from
A to G? Give your answer to 1 decimal place. (3 marks, ★★★★★)

NAILIT!

When solving 3D
problems, sketch the
right-angled triangle
you are using on
its own.

.... cm

3) A, B, C and D are points on the circumference of a circle with centre O
and diameter 10.8 cm.

$CD = 5.8$ cm and angle $BAC = 65°$.

a Calculate the length AD. Give your answer
to 2 decimal places. (2 marks, ★★★★)

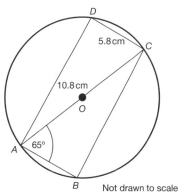

Not drawn to scale

.... cm

b Calculate the area of the quadrilateral $ABCD$. Give your answer to 2 decimal places.
(4 marks, ★★★★★)

.... cm^2

[Total: 6 marks]

Area of 2D shapes

① *ABCD* is a rectangle. (★★★★)

   E and *F* are points on *AB* such that *DE* = 9 cm and *CF* = 8 cm.

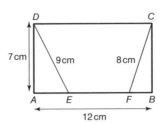

NAILIT!

You need to know the formulae for the areas of a triangle, a parallelogram, a trapezium and a circle.

  a **i** Calculate the length of *AE*. Give your answer to 2 decimal places. (2 marks)

    ii Hence find the area of triangle *ADE*. Give your answer to 2 decimal places. (1 mark)

.... cm

.... cm²

  b Calculate the area of *CDEF* giving your answer to 2 decimal places. (3 marks)

.... cm² **[Total: 6 marks]**

② Two shapes *ABC* and *DEFG* are shown. The shapes have the same perimeter. The measurements for both are in centimetres. (★★★)

  a Work out the area of shape *DEFG*. (4 marks)

.... cm²

  b Work out the area of shape *ABC*. (2 marks)

.... cm²

Not drawn to scale

[Total: 6 marks]

③ A stencil for a border pattern is made of semicircles with radius 3 cm, arranged as shown in the diagram.

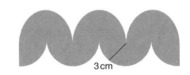

3 cm

  a Work out the area of the pattern. Give your answer in terms of π. (2 marks, ★★★★)

.... cm²

  b Work out the perimeter of the pattern. Give your answer in terms of π. (2 marks, ★★★)

.... cm²

[Total: 4 marks]

Volume and surface area of 3D shapes

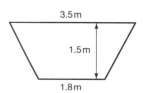

3.5 m

1.5 m

1.8 m

(1) The cross-section of a skip is in the shape of a trapezium. (★★★)

a Calculate the cross-sectional area of the skip shown in the above diagram.
(2 marks)

b The skip is a prism with the cross-sectional area shown and a length of 1.6 m. Calculate the volume of the skip. (2 marks)

> The volume of a prism is the area of cross-section × length.

.... m²

.... m³

[Total: 4 marks]

(2) A glass is in the shape of a hemisphere of radius 4 cm.

How many of these glasses can be filled from a 750 ml bottle? (5 marks, ★★★★★)

....

(3) **a** A solid cone has base radius 4.8 cm and perpendicular height 5.6 cm.

Work out the slant height of the cone. Give your answer to 1 decimal place. (2 marks, ★★★★)

> Slant height is the distance from the tip to the circular edge. Use Pythagoras' theorem to calculate it.

.... cm

b A sphere has the same total surface area as the total surface area of the cone.

Work out the radius of the sphere. Give your answer to 1 decimal place. (4 marks, ★★★★★)

SNAPIT!

Surface area of sphere $= 4\pi r^2$
Curved surface area of cone $= \pi r l$
Volume of a cone $= \frac{1}{3}\pi r^2 h$
Volume of a cylinder $= \pi r^2 h$

.... cm **[Total: 6 marks]**

(4) The diagram shows a cone with a radius of 2 cm and height 12 cm. The cone is filled to the top with water. All of the water is then poured into a cylinder with radius 5 cm.

Work out the depth of the water in the cylinder. (4 marks, ★★★★)

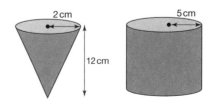

2 cm

5 cm

12 cm

.... cm

Trigonometric ratios

NAILIT!

Learn a mnemonic (e.g. SOH CAH TOA) to remember the ratios.

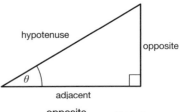

$\sin\theta = \dfrac{\text{opposite}}{\text{hypotenuse}}$ (SOH)

$\cos\theta = \dfrac{\text{adjacent}}{\text{hypotenuse}}$ (CAH)

$\tan\theta = \dfrac{\text{opposite}}{\text{adjacent}}$ (TOA)

(1) The diagrams show two right-angled triangles. (★★)

a Calculate the length of the side marked x.
Give your answer to 1 decimal place. (2 marks)

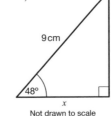

.... cm

b Calculate the size of angle y. (2 marks)

.... °

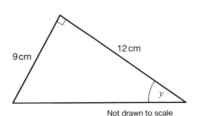

Start by naming the sides and then choose the correct trigonometric function (sin, cos or tan).

[Total: 4 marks]

(2) *ABC* is a right-angled triangle.

$AB = 1.5\,\text{m}$, $BC = 2.1\,\text{m}$ and angle $BAC = 90°$. ⟵ First draw a sketch diagram.

Work out the size of angle *ABC*. Give your answer to 1 decimal place. (2 marks, ★★)

.... °

(3) Triangle *ABC* is an isosceles triangle.

$AC = BC = 10\,\text{cm}$.

Calculate the perpendicular height, h, of the triangle.

Give your answer to 2 decimal places.
(3 marks, ★★★)

NAILIT!

You often split isosceles triangles into two right-angled triangles and then use either Pythagoras' theorem or trigonometry.

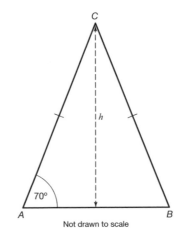

.... cm

4 The diagram shows a cuboid with length 5.5 cm, height 2.5 cm and depth 3 cm.

Work out the angle between the longest diagonal and the base of the box. (3 marks, ★★★★★)

NAILIT!

In 3D questions, sketch the right-angled triangles that you need. Then use Pythagoras' theorem or trigonometry to work out what you are asked to find.

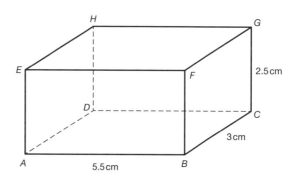

.... °

STRETCHIT!

In the diagram above, *HB* would be an internal diagonal. Can you find the lines that are the other internal diagonals? How many are there in all?

HB is in a right-angled triangle, *HDB*. Write down the right-angled triangles for the other diagonals that you found.

Exact values of sin, cos and tan

NAILIT!

You need to remember or be able to work out exact values of sin, cos and tan of 0°, 30°, 45°, 60° and the sin and cos of 90°. Remember that exact values of trigonometric expressions involve only whole numbers, fractions or surds.

① Show that

$$\tan 45° + \cos 60° = \frac{3}{2}$$ (3 marks, ★★★★)

② a Write the exact value of (★★★)

    i sin 45° (1 mark) ◄— Draw a sketch if you need to. When asked for the exact value, use surds.

    ....

    ii cos 45° (1 mark)

    ....

b Hence prove that $\frac{\sin 45°}{\cos 45°} = \tan 45°$

(3 marks, ★★★★)

[Total: 5 marks]

③ The diagram shows a right-angled triangle *ABC*. (★★★)

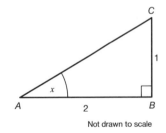

Not drawn to scale

a Find the length of *AC*. Give your answer as a surd. (2 marks)

....

b Write down the exact values of

    i sin *x* (1 mark)

    ....

    ii cos *x* (1 mark)

    ....

c Hence show that $(\sin x)^2 + (\cos x)^2 = 1$

(3 marks)

[Total: 7 marks]

④ Without using a calculator, show that $\tan 30° + \tan 60° + \cos 30° = \frac{11\sqrt{3}}{6}$ (3 marks, ★★★★★)

Sectors of circles

SNAP IT!

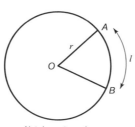

Not drawn to scale

(1) The diagram shows a circle with centre O and radius r.

The length of the arc $l = 3$ cm and the radius $r = 5$ cm.

Find angle AOB. Give your answer to the nearest degree. (2 marks, ★★★)

You need to remember these formulae:

- circumference $= 2\pi r$
- area of circle $= \pi r^2$
- length of arc $= \frac{\theta}{360} \times 2\pi r$
- area of sector $= \frac{\theta}{360} \times \pi r^2$

.... °

Use the formula for the length of an arc. Make sure you calculate the area of the major sector, not the minor sector.

(2) The diagram shows a circle with radius 12 cm and angle $AOB = 70°$.

Calculate the shaded area. Give your answer to 1 decimal place. (3 marks, ★★★)

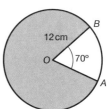

Not drawn to scale

.... cm^2

(3) These two sectors have the same area. (★★★★)

a Work out the radius of sector B.
Give your answer to 3 significant figures. (3 marks)

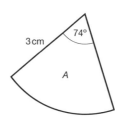
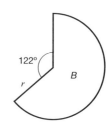

Not drawn to scale

.... cm

b Work out the ratio of the perimeter of sector A to the perimeter of sector B in the form $1:n$. Give n to 2 decimal places. (3 marks)

....

[Total: 6 marks]

(4) The diagram shows a circle with centre O and radius 6 cm. AB is a chord of the circle. The length of the arc AB is 5.4 cm. (★★★★★)

a Find the area of the minor sector AOB. Give your answer to 1 decimal place. (3 marks)

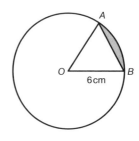

....

b Find the area of the shaded segment. Give your answer to 1 decimal place. (3 marks)

....

[Total: 6 marks]

Sine and cosine rules

SNAP IT!

(1) *ABC* is a triangle with sides
AB = 25 cm, *BC* = 30 cm
and acute angle *ABC* = θ,
where sin θ = $\frac{3}{5}$. (★★★★★)

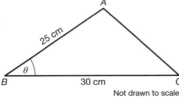

Not drawn to scale

a Calculate the area of triangle *ABC*. (2 marks)

.... cm²

b Find the exact value of cos θ. (2 marks)

....

c Calculate the length of *AC*. Give your answer to 3 significant figures. (3 marks)

.... cm

[Total: 7 marks]

You need to know these formulae.

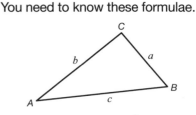

Sine rule: $\dfrac{a}{\sin A} = \dfrac{b}{\sin B} = \dfrac{c}{\sin C}$

Cosine rule: $a^2 = b^2 + c^2 - 2bc\cos A$

Area: $\frac{1}{2}ab\sin C$

(2) The acute-angled triangle *ABC* is shown. (★★★★)

a Find *x* to three significant figures. (2 marks)

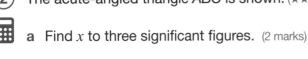

Notice the pairing of sides and angles. This indicates that you need to use the sine rule.

.... cm

b Hence find the area of the triangle.
Give your answer to 3 significant figures. (2 marks)

.... cm²

[Total: 4 marks]

(3) In triangle *ABC*, *AB* = 4 cm, *BC* = (3√2 − 1) cm and angle *BAC* = 30°.
Find an expression for the sine of the angle *ACB* in the form $\dfrac{2 + m\sqrt{2}}{n}$,
where *m* and *n* are integers. (4 marks, ★★★★★)

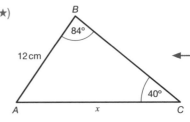

Draw yourself a sketch diagram first.

....

Vectors

WORKIT!

In triangle OAB, $\overrightarrow{OA} = \mathbf{a}$ and $\overrightarrow{OB} = \mathbf{b}$.

M is the midpoint of AB.

a Write vector $\overrightarrow{BM}$ in terms of $\mathbf{a}$ and $\mathbf{b}$.

$\overrightarrow{AB} = \overrightarrow{AO} + \overrightarrow{OB} = -\mathbf{a} + \mathbf{b} = \mathbf{b} - \mathbf{a}$

$\overrightarrow{BM} = \frac{1}{2}\overrightarrow{BA} = \frac{1}{2}(-\mathbf{b} + \mathbf{a}) = \frac{1}{2}(\mathbf{a} - \mathbf{b})$

b Find the vector $\overrightarrow{BM}$ if $\mathbf{a} = \begin{pmatrix} 5 \\ 2 \end{pmatrix}$ and $\mathbf{b} = \begin{pmatrix} 6 \\ 0 \end{pmatrix}$

$\overrightarrow{BM} = \frac{1}{2}(\mathbf{a} - \mathbf{b}) = \frac{1}{2}\left(\begin{pmatrix} 5 \\ 2 \end{pmatrix} - \begin{pmatrix} 6 \\ 0 \end{pmatrix}\right) = \frac{1}{2}\begin{pmatrix} -1 \\ 2 \end{pmatrix}\begin{pmatrix} \frac{-1}{2} \\ 1 \end{pmatrix}$

NAILIT!

If two vectors are parallel they have an identical vector part:

$2\mathbf{a} - \mathbf{b}$ is parallel to $\mathbf{a} - \frac{\mathbf{b}}{2}$

as it can be written as $2\left(\mathbf{a} - \frac{\mathbf{b}}{2}\right)$.

It is twice the length of $\mathbf{a} - \frac{\mathbf{b}}{2}$.

---

① $\mathbf{p} = \begin{pmatrix} -2 \\ 3 \end{pmatrix}$ and $\mathbf{q} = \begin{pmatrix} 6 \\ -1 \end{pmatrix}$.

Work out as a column vector (★★★)

a $\frac{1}{2}(\mathbf{p} + \mathbf{q})$ (2 marks)

b $2\mathbf{p} - 3\mathbf{q}$ (2 marks)

....

....

[Total: 4 marks]

② OAB is a triangle, $\overrightarrow{OA} = \mathbf{a}$ and $\overrightarrow{OB} = \mathbf{b}$. (★★★★)

a Find $\overrightarrow{AB}$ in terms of $\mathbf{a}$ and $\mathbf{b}$. (2 marks)

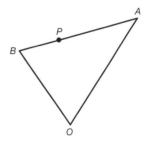

....

b P is the point on AB such that $AP:BP = 3:2$.
Find $\overrightarrow{AP}$ in terms of $\mathbf{a}$ and $\mathbf{b}$. (3 marks) ◄——— Line AB is split into $3 + 2 = 5$ parts.

....

c Q is a point on OA such that $OQ:QA = 2:3$. Are lines QP and OB parallel? Explain your answer. (2 marks)

....

[Total: 7 marks]

(3) *ABC* is a triangle and *ACD* is a straight line.

$\overrightarrow{AC} = \mathbf{a}$ and $\overrightarrow{AB} = 3\mathbf{b}$.

P is a point on *AB* such that $AP:PB = 2:1$.

M is the midpoint of *BC*.

C is the midpoint of *AD*. (★★★★★)

a Find $\overrightarrow{BC}$ in terms of **a** and **b**. (1 mark)

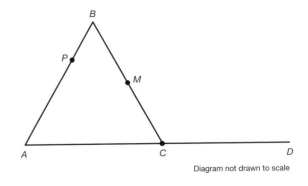

Diagram not drawn to scale

b Show that *PMD* is a straight line. (4 marks)

[Total: 5 marks]

Probability
The basics of probability

SNAP IT!

If all outcomes are equally likely

Probability $= \dfrac{\text{number of ways something can happen}}{\text{total number of possible outcomes}}$

This is a mathematical calculation and is known as **theoretical probability**.

(1) Two dice are thrown and the scores are added together. (★★)

The sample space diagram for the total score is shown below. The first row has been completed.

a Complete the sample space diagram.
(2 marks)

Dice 1

	1	2	3	4	5	6
1	2	3	4	5	6	7
2						
3						
4						
5						
6						

Dice 2

b Find the probability of obtaining a score of 12. (1 mark)

....

c Find the probability of obtaining a score that is a prime number. (1 mark)

....

d What is the most likely score? Give a reason for your answer. (1 mark)

....

....

[Total: 5 marks]

(2) A box of chocolates contains three different types of chocolate.

The table shows the number of chocolates of each type.

	Truffle	Mint	Caramel
Number of chocolates	$2x + 1$	x	$2x$

You need to find the value of x first.

A chocolate is chosen at random. The probability of the chocolate being a mint is $\frac{4}{21}$. (★★★)

a Calculate the total number of chocolates in the box. (2 marks)

b Calculate the probability of choosing a truffle. (1 mark)

....

....

[Total: 3 marks]

(3) Amy and Bethany each throw a fair dice.

Calculate the probability that the score on Amy's dice is (★★★★)

a equal to the score on Bethany's dice (2 marks)

....

b greater than the score on Bethany's dice. (2 marks)

....

[Total: 4 marks]

Probability experiments

50 light bulbs were tested; 44 lasted for 1500 hours or more, and 6 for less than 1500 hours.

In a batch of 1000 light bulbs, how many would be expected to last for less than 1500 hours?

Relative frequency for less than 1500 hours

$= \frac{6}{50}$, so frequency $= \frac{6}{50} \times 1000$

$= 120$ light bulbs

NAILIT!

When probability experiments are conducted (e.g. throwing a dice, spinning a spinner, etc.)

Relative frequency $= \frac{\text{frequency of particular event}}{\text{total trials in experiment}}$

The value for the relative frequency will only approach the theoretical probability over a very large number of trials.

1. A pentagonal unbiased spinner with sides numbered 1 to 5 was spun 100 times. (★★★)

The frequency that the spinner landed on each number was recorded in the table below.

Score on spinner	1	2	3	4	5
Frequency	18	23	22	19	18

a Abdul says that the spinner must be biased because if it was fair the frequency for each score would be the same. Explain why Abdul is wrong. (1 mark)

....

....

b Calculate the relative frequency of obtaining a score of 3 on the spinner. Give your answer as a fraction in its simplest form. (2 marks)

....

c If the spinner was spun 500 times, use the relative frequency to estimate how many times the spinner would give a score of 4. (1 mark)

....

[Total: 4 marks]

(2) A hexagonal spinner with sides numbered from 1 to 6 has the
following relative frequencies of scores from 1 to 6. (★★★)

Score	1	2	3	4	5	6
Relative frequency	$3x$	0.05	$2x$	0.25	0.2	0.1

a Calculate the value of x. (2 marks)

....

b Calculate the relative frequency of obtaining a score of 1. (1 mark)

....

c The spinner was spun 80 times. Estimate how many times the score was 5. (1 mark)

....

[Total: 4 marks]

 **STRETCHIT!**

Playing cards are interesting because
they offer a lot of different probabilities.
If you draw a card, it could be: red,
black, a heart, a diamond, a club, a
spade, an ace, a picture card, a 5 and
so on. Think about how these different
probabilities affect the games people
play – and conjuring tricks too.

The AND and OR rules

SNAPIT!

The OR rule: Probability of A or B happening: P(A OR B) = P(A) + P(B)

The AND rule: Probability of A and B happening: P(A AND B) = P(A) × P(B)

① A card is chosen at random from a pack of 52 playing cards and noted. The card is returned, the pack shuffled and a second card chosen. (★★★)

Find the probability the two cards were

a two picture cards (1 mark)

c the queen of diamonds and the queen of hearts. (1 mark)

....

....

b an ace and a picture card (1 mark)

[Total: 3 marks]

....

② A bag contains 10 marbles: 3 red, 5 blue and 2 green.

A marble is removed from the bag and its colour noted before it is put back into the bag.

Another marble is removed and its colour is also noted. (★★★★)

a Explain what is meant by independent events. (1 mark)

....

b Calculate the probability that two red marbles were picked. (2 marks)

c Calculate the probability that the two marbles were red and blue. (3 marks)

....

....

[Total: 6 marks]

③ The probability that Aisha is given maths homework on a certain day is $\frac{3}{5}$. The probability she is given French homework is $\frac{3}{7}$. The probability she is given geography homework is $\frac{1}{4}$. (★★★★)

a Work out the probability that she is given homework in all three subjects.

(2 marks)

b Work out the probability that she is not given homework in any of these subjects.

(2 marks)

....

....

[Total: 4 marks]

Tree diagrams

WORKIT!

A box contains 7 counters: 3 red and 4 blue.

A counter taken at random from the box is red.

A second counter is taken at random.

What is the probability that the second counter is red if

a the first counter was put back in the box

Number of red counters in box = 3

Total number of counters = 7

$P(red) = \frac{3}{7}$

b the first counter was not put back in the box?

Number of red counters in box = 2

Total number of counters = 6

$P(red) = \frac{2}{6} = \frac{1}{3}$

NAILIT!

In independent events, the probabilities do not affect each other. In conditional events, the probability of the second outcome depends on what the first outcome was.

---

(1) From a group of children consisting of 4 girls and 6 boys, 2 children are picked at random to take part in an interview.

Work out the probability that (★★★★)

a two girls are chosen (3 marks)

NAILIT!

Draw a tree diagram showing the possible selections.
The probability of a particular sequence is found by multiplying the probabilities along the branches making up the sequence.

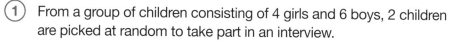

b a boy and a girl are chosen. (3 marks)

[Total: 6 marks]

② There are 450 pupils in a junior school, 56% of whom are boys.

The ratio of boys who have school dinners to the boys who have a packed lunch is 2:1. The ratio of girls who have school dinners to the girls who have a packed lunch is 5:4 (★★★★)

> Make sure you understand the difference between probability trees and frequency trees.

a Complete the frequency tree. (4 marks)

b Using the frequency tree, work out the probability that a pupil chosen at random from the school is a girl who has a school dinner. (2 marks)

....

c Using the frequency tree, work out the probability that a pupil chosen at random from the school has a school dinner. (2 marks)

> There are two ways in which this can happen so you need to find the probability for each and then add them together.

....

[Total: 8 marks]

③ A bag contains 9 marbles, of which 5 are red, 3 are green and 1 is yellow. Three marbles are chosen at random from the bag.

Giving your answer correct to 3 decimal places, calculate the probability of choosing (★★★★★)

a 1 marble of each colour (3 marks)

> Draw a probability tree to help you.

....

b no green marbles (2 marks)

....

c 3 marbles of the same colour. (3 marks)

....

[Total: 8 marks]

Venn diagrams and probability

SNAPIT!

In Venn diagrams, the whole rectangle represents everything being considered and is called the universal set, ξ

You need to be familiar with set notation:

- $A \cap B$ (A intersect B): all the elements in set A that are also in set B
- $A \cup B$ (A union B): all the elements in sets A and B combined
- A' (A complement): all the elements in the universal set that are not in set A.

(1) The Venn diagram shows the universal set with two subsets A and B. (★★★)

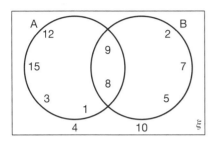

Write down the numbers in

a $A \cap B$ (1 mark)

9, 8

b $A \cup B$ (1 mark)

1, 3, 12, 15, 2, 7, 5
+ 8 8 9

c B' (1 mark)

1, 3, 4, 10, 12, 14

d $(A \cup B)'$ (1 mark)

10, 4

[Total: 4 marks]

(2) $\xi = \{1, 2, 3, 4, 5, 6, 7, 8, 9, 10, 11, 12, 13, 14, 15\}$

P = prime numbers

O = odd numbers (★★★★)

a Complete the Venn diagram. (3 marks)

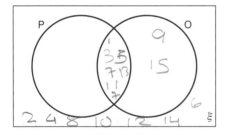

b One of the numbers in the universal set is chosen at random.

Calculate the probability that the number chosen is in $P \cap O$. (2 marks)

[Total: 5 marks]

(3) A survey asked 100 sports students about the types of sport they played: individual, small team, large team. It found that 6 students played individual, small team and large team sports; 15 only played individual sports; 10 played both individual and small team sports; and 18 played both small team and large team sports. Overall, there were 41 who played individual sports and 30 who played small team sports. (★★★★★)

a A sports student is chosen at random. What is the probability that the student plays team sports? (4 marks)

b A student is chosen at random from those students who play large team sports. What is the probability that they also play small team sports? (2 marks)

[Total: 6 marks]

Statistics
Sampling

(1) Dorota and Ling are going to perform a survey at two different schools. Both schools have 900 students. Dorota intends to take a sample of 50 students and Ling intends to take a sample of 100 students. (★★★)

   a Whose survey is likely to be more reliable and why? (2 marks)

....

....

   b There are 500 girls and 400 boys in Dorota's school. How many boys should she include in her sample? (2 marks)

> The proportion of boys in the sample should be the same as the proportion in the school.

....

[Total: 4 marks]

(2) A survey is to be taken about parking in a town centre. There are 1000 parking spaces in the town, and a sample of people who use them is to be used for the survey.

State two ways of making sure that the sample is unbiased. (2 marks, ★★)

....

....

....

....

(3) A health survey is being carried out to look at peoples' opinions about smoking.

It has been found that 52% of the population is male and that 12% of them smoke. Of the females, 10% smoke.

> You may use a calculator for these questions.

A stratified sample of 400 people is to be used. Each person in the sample is to be given a questionnaire to fill in. (★★★)

   a How many questionnaires should be given to females who do not smoke? (2 marks)

....

   b How many questionnaires should be given to males who smoke? (2 marks)

....

[Total: 4 marks]

Two-way tables, pie charts and stem-and-leaf diagrams

(1) The two-way table gives information about the favourite TV soap of a sample of 100 students in a college. (★★★)

	Coronation Street	EastEnders	Emmerdale	Total
Boys	12	31	20	
Girls		12		37
Total	30			

a Complete the two-way table. (3 marks)

b A student is picked at random from the college. Find the probability that the student's favourite TV soap is Emmerdale. (1 mark)

....

c A girl is picked at random. Find the probability that her favourite TV soap is EastEnders. (1 mark)

....

[Total: 5 marks]

NAILIT!

In a two-way table, the numbers must add up across the rows and down the columns to give the totals.

(2) The table gives information about eye colour in a secondary school class.

Complete the pie chart to show this information. (4 marks, ★★)

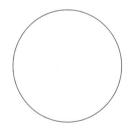

Colour	Frequency
Blue	11
Green	3
Brown	16

Work out the angle that represents one person. Then find the angle for each of the frequencies in the table.

(3) The headteacher of a school recorded the number of students who were late each day over a 21-day period. Here are the results. (★★)

12	8	10	22	20	17	12
5	12	18	11	8	16	4
10	12	11	3	4	9	26

a Draw a stem-and-leaf diagram to show this information. (2 marks)

Don't forget you must include a key.

b Find the median number of students late each day during this period. (1 mark)

c Find the range of the number of students who were late. (1 mark)

....

.... **[Total: 4 marks]**

Line graphs for time series data

(1) The table shows sales of barbecues at a garden centre for the first 6 months of the year. (★★★★)

Month	Jan	Feb	Mar	Apr	May	June
Sales	12	10	20	54	87	130

a Work out the 3-point moving averages for this data. You may use a calculator. (4 marks)

January to March: ....

February to April: ....

March to May: ....

April to June: ....

b Use your answers in part a to describe the trend in sales of barbecues over the period. (1 mark)

....

[Total: 5 marks]

NAILIT!

Time series data shows seasonal peaks and troughs, so it is hard to see trends. Moving averages smooth out the data, so trends are easier to spot.

WORKIT!

A shop's sales over 5 years are shown.

Year	Year 1	Year 2	Year 3	Year 4	Year 5
Sales (£ thousands)	53	48	54	67	62

Work out the 3-point moving average.

Years 1 to 3: $\frac{53 + 48 + 54}{3} = 51.7$

Years 2 to 4: $\frac{48 + 54 + 67}{3} = 56.3$

Years 3 to 5: $\frac{54 + 67 + 62}{3} = 61$

(2) This table shows sales of a popular toy in a shop. The sales seem to be seasonal as they show a pattern that repeats in both years. (★★★★)

Year	Quarter	Sales
2015	1	114
	2	142
	3	155
	4	136
2016	1	116
	2	150
	3	153
	4	140

a Plot the values using a 4-point moving average. (4 marks)

b Describe the trend shown by the graph. (1 mark)

....

....

[Total: 5 marks]

The moving average value is plotted at the centre of each time interval being used.

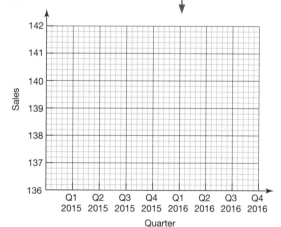

Averages and spread

(1) The number of road accidents along a certain stretch of road was recorded each month over a year. The results were (★)

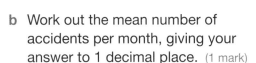

5   3   1   4   3   5   0   1   4   1   2   3

a Write down the total number of accidents over the year. (1 mark)

> You may use a calculator for these questions.

....

b Work out the mean number of accidents per month, giving your answer to 1 decimal place. (1 mark)

c Write down the median number of accidents over the year. (1 mark)

....   **[Total: 3 marks]**

(2) The mean mark of one group of 25 students is 60. The mean mark for a different group of 30 students is 72. Calculate the mean mark for the combined group of 55 students. Give your answer to 1 decimal place. (3 marks, ★★★)

> You need to find the total marks awarded to both groups of students, then divide by the total number of students.

....

(3) A vet records the ages of all the dogs she sees in a week.

The table summarises the information. (★★★)

Age (t years)	Frequency	Mid-interval value	Frequency × mid-interval value
$0 < t \leq 4$	8		
$4 < t \leq 8$	10		
$8 < t \leq 12$	16		
$12 < t \leq 14$	1		

> You may use a calculator for these questions.

a Complete the table. (3 marks)

b Write down how many dogs she saw at her practice that week. (1 mark)

c Estimate the mean age of the dogs she saw, giving your answer to 2 significant figures. (2 marks)

.... years

[Total: 6 marks]

(1) The table gives information about the ages of children at a nursery.

Age (*a* years)	Frequency
$0 < a \leq 0.5$	8
$0.5 < a \leq 1$	12
$1 < a \leq 2$	16
$2 < a \leq 3$	30
$3 < a \leq 5$	28

Draw a histogram for the information shown in the table. (5 marks, ★★★★)

NAILIT!

In histograms with class intervals of different widths, frequency density is plotted on the vertical axis. The area of each bar, not the height, represents the frequency.

Frequency (area of bar) = class width (width) × frequency density (height)

> Work out the frequency density for each class.

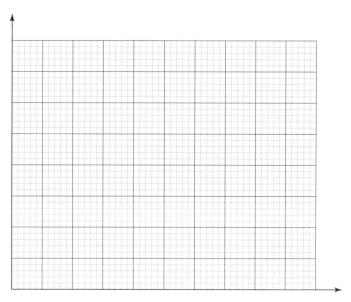

(2) The table and histogram show information about the wingspan of canaries in an aviary.

Use the histogram to complete the table. (4 marks, ★★★★★)

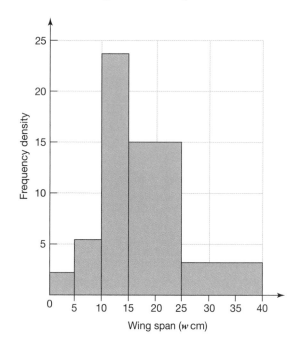

Wingspan (*w* cm)	Frequency
$0 < w \leq 5$	
$5 < w \leq 10$	6
$10 < w \leq 15$	24
$15 < w \leq 25$	30
$25 < w \leq 40$	

Cumulative frequency graphs

(1) Tom receives a lot of junk mail through the post. He decides to keep a record of the amount of unwanted mail he receives over a period of two months. The table records the number of items per day and their corresponding frequency. (★★★)

NAILIT!

Cumulative frequency graphs have cumulative frequency (running total of the frequency) plotted against the upper value of each class interval.

Number of items of junk mail per day (n)	Frequency (days)	Cumulative frequency
$0 < n \leq 2$	21	
$2 < n \leq 4$	18	
$4 < n \leq 6$	10	
$6 < n \leq 8$	8	
$8 < n \leq 10$	3	
$10 < n \leq 12$	1	

a Write down the modal class interval. (1 mark)

....

b Complete the cumulative frequency column. (2 marks) ◀ This is the running total.

c On the grid, draw the cumulative frequency graph for the data in the table. (3 marks) ◀ Plot the upper value of the class interval against its cumulative frequency.

d Use your graph to find the median daily number of items of junk mail Tom received over the two months. (1 mark)

....

[Total: 7 marks]

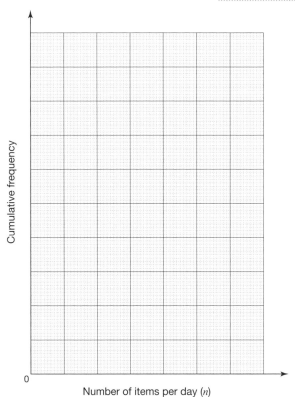

Cumulative frequency

0

Number of items per day (n)

② The cumulative frequency graph gives information about the masses of young emperor penguins in a zoo. (★★★★★)

a **i** Use the graph to find the median mass. (1 mark)

> You may use a ruler to draw lines on the graph to help you.

.... kg

ii Find the interquartile range of the masses. (2 marks)

.... kg

b Estimate the number of penguins with a mass above 10 kg.
Show your workings. (2 marks)

....

[Total: 5 marks]

STRETCHIT!

The graph below has a point (25, 40).

Ruby says, 'What big penguins they've got at the zoo – 40 of them weigh 25 kg!'

Ruby is interpreting the graph incorrectly. Make sure you understand why.

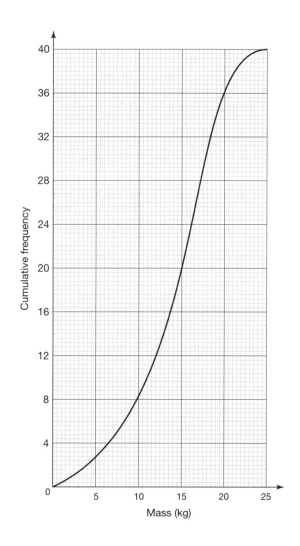

Comparing sets of data

(1) Here are Chloe's scores in 15 spelling tests. (★★★)

10 12 17 17 15 18 17 16 14 15 14 12 17 19 18

NAILIT!

When comparing sets of data compare both an average (mean, median or mode) and a measure of spread (range or interquartile range).

a On the grid, draw a box plot for this information. (3 marks)

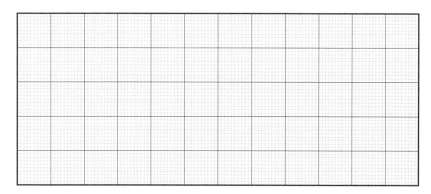

Sasha takes the same 15 spelling tests and her results are summarised below.

Her median score is 17.

The interquartile range for her scores is 2.

The range of her scores is 5.

b Compare Sasha and Chloe's scores. (2 marks)

....

....

....

....

....

....

....

[Total: 5 marks]

(2) There are 100 male runners taking part in a race. The cumulative frequency graph of their race times is shown below. (★★★★★)

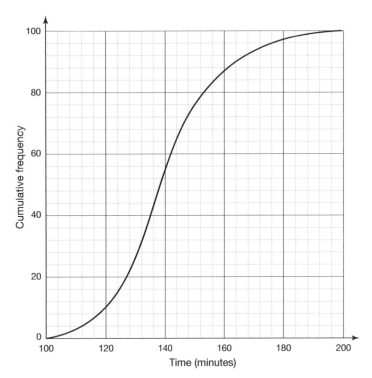

a Use the graph to work out the following for the male runners

  i the median time (1 mark)

  .... minutes

  ii the range (1 mark)

  .... minutes

  iii the interquartile range (1 mark)

  .... minutes

b Here is a box plot for 100 female runners who ran in the same race. (4 marks)

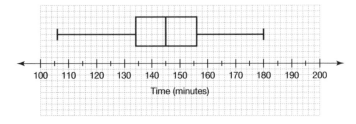

Compare the performance of the male and female runners.

....

....

....

[Total: 7 marks]

Scatter graphs

NAILIT!

Scatter graphs are used to investigate whether a correlation exists between two quantities. If the points show an upward slope to the right there is a positive correlation and if they show a downward slope to the right there is a negative correlation. The **line of best fit** demonstrates this. When drawing one, position it so that the points are evenly distributed on both sides.

(1) A driving school advertises for students. The manager would like find out how the amount spent on advertising affects the number of new students booking lessons as a result of the adverts. She has collected the data shown in the table. (★★)

Advertising spend per week (£)	Number of new students per week from adverts
20	1
30	2
40	4
50	4
60	5
70	6

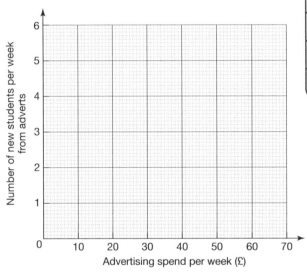

a Using the grid, draw a scatter graph to show this information. (3 marks)

b Draw a line of best fit on your graph. (1 mark)

c Describe the type of correlation. (1 mark)

....

[Total: 5 marks]

(2) The scatter graph below shows the test marks in physics and maths for the same set of students. (★★)

a Draw a circle around the point that is an outlier.
(1 mark)

b Ignoring the outlier, draw the line of best fit.
(1 mark)

c A student obtained a mark of 70% in the physics exam but was absent for the maths exam. Use your graph to predict the student's likely mark in the maths exam.
(1 mark)

....

[Total: 3 marks]

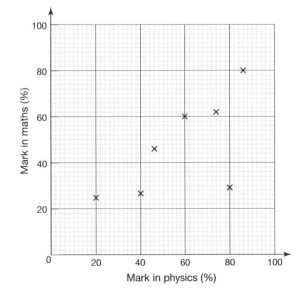

Practice paper (non-calculator)

Higher tier

Time: 1 hour 30 minutes

The total mark for this paper is 80.
The marks for **each** question are shown in brackets.

1   Work out 3.54×8.3.

                                                      [Total: 2 marks]

2   a   Work out $1\frac{2}{5} \div \frac{21}{50}$.

                                                      [2 marks]

       b   Work out $2\frac{7}{8} - \frac{3}{5}$.

                                                      [2 marks]
                                                    [Total: 4 marks]

3      $486 \times 29 = 14\,094$

     Use this fact to find the value of

     a 0.486×2.9

                                     1.4094
                                                  [1 mark]

     b $140.94 \div 29$

                                     4.86
                                                  [1 mark]
                                                    [Total: 2 marks]

4   Expand and simplify $(a + 3)(a + 9)$.

$$a^2 + 9a + 3a + 27$$

$$a^2 + 12a + 27$$

[Total: 2 marks]

5   Factorise fully $4y^2 - 64$.

2

$(2y - 8)(2y + 8)$

$(2y - 8)(2y + 8)$

[Total: 2 marks]

6   On a farm, there are 5 times as many sheep as cattle.

There are 3 times as many cattle as chickens.

Write down the ratio of the number of sheep to the number of cattle to the number of chickens.

[Total: 3 marks]

7   Simplify

a $(2x^2y)^3$

[1 mark]

b $2x^{-3} \times 3x^4$

[1 mark]

c $\dfrac{15a^3b}{3a^3b^2}$

$5b$

$5b$

[1 mark]

[Total: 3 marks]

8   **a** $x^2 - 2x + 6$ can be written in the form $(x - a)^2 + b$.

         Find the values of the integers a and b.

....

[2 marks]

     b **i** Write down the coordinates of the turning point of the curve with equation

           $y = x^2 - 2x + 6$.

....

[1 mark]

         ii Hence draw a sketch graph of the curve with equation $y = x^2 - 2x + 6$. On your graph mark the coordinates of the turning point and the point where the curve cuts the y-axis.

[2 marks]

         iii Use your graph to explain why the equation $y = x^2 - 2x + 6$ has no roots.

....

[1 mark]

[Total: 6 marks]

9   **a** Write down the value of $100^{\frac{1}{2}}$.

.....................10.....................

[1 mark]

     b Find the value of $\left(\dfrac{125}{64}\right)^{-\frac{2}{3}}$.

$$\left(\frac{64}{125}\right)^{\frac{3}{2}} \qquad \sqrt[3]{\frac{64}{125}} \quad \text{and then} \atop \text{square}$$

....

[2 marks]

[Total: 3 marks]

10   Make y the subject of $ay + c = \dfrac{3 - y}{2}$.

....

[Total: 3 marks]

11   Two metal spheres are mathematically similar. The ratio of their surface areas is $25:49$.
        The volume of the large sphere is $1000\,\text{cm}^3$.

        Find the volume of the small sphere.
        Give your answer to the nearest integer.

.... cm^3

[Total: 3 marks]

12   Show that $\dfrac{\sqrt{3} - 2}{\sqrt{3} + 1}$ can be written as $\dfrac{5 - 3\sqrt{3}}{2}$.

[Total: 3 marks]

13   Show that $0.1\dot{3}\dot{6}$ can be written as the fraction $\dfrac{3}{22}$.

$1000x = 136.\dot{3}\dot{6}$

$100x = 1.\dot{3}\dot{6}$

$990x = 135 \quad \rightarrow \dfrac{135}{990} \quad \rightarrow \dfrac{15}{110} \quad \xrightarrow{\div 5} \dfrac{3}{22}$

[Total: 3 marks]

311

14   Below is a sketch of the graph of $y = f(x)$.

The graph has a turning point at (0, 6).

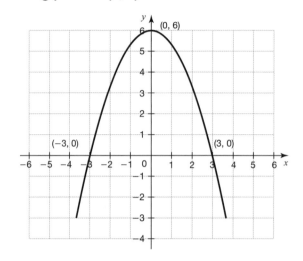

Write down the coordinates of the turning point of the curve with equation

a $y = -f(x)$

....

[1 mark]

b $y = f(-x)$

....

[1 mark]

[Total: 2 marks]

15

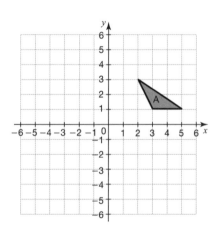

a   Enlarge shape **A** by scale factor −1 and centre of enlargement (2, −1).

Label your image **B**.

[3 marks]

b   Describe fully a different transformation that will map shape **A** onto shape **B**.

....

[1 mark]

[Total: 4 marks]

16   Solve $x^2 \le 2x + 15$.

....

[Total: 3 marks]

17   a Using the axes shown below, sketch the graph of $y = \tan\theta$ for values of θ in the
        range $0° \le \theta \le 360°$.

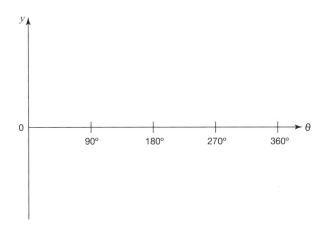

[2 marks]

        b Using the graph you sketched in part a to help you, solve the equation

        $\tan\theta = 1$ giving all values of θ in the range $0° \le \theta \le 360°$.

....

[2 marks]

[Total: 4 marks]

18   *OA* is a straight line passing through the origin and point *A*(3, 4). A straight line *BC*
        passes through A and is perpendicular to line *OA*.

        Find the equation of line *BC*.

        Give your answer in the form $ax + by + c = 0$ where a, b and c are integers.

....

[Total: 4 marks]

19   Solve the simultaneous equations

$6x + 5y = 35$

$x - 2y = 3$

[Total: 4 marks]

20   Find the nth term for the sequence

5, 12, 25, 44, ...

7   13   19

+6   +6

$\frac{6}{3} = 3n^2$

n	1	2	3	4
$3n^2$	3	12	27	48
term	5	12	25	44

[Total: 5 marks]

21   The speed–time graph below shows the motion of a particle.

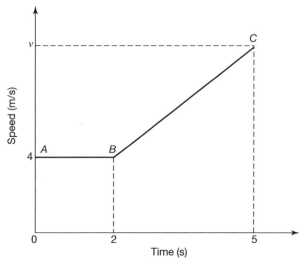

[Total:]

a Describe the motion between points

   i *A* and *B*

   ....

   [1 mark]

   ii *B* and *C*.

   ....

   [1 mark]

b The total distance travelled by the particle between *A* and *C* is 29 m.

   Find the value of speed v.

   .... m/s

   [2 marks]

   [Total: 4 marks]

22   The table shows the distribution of marks for 98 students in a physics test.

Mark (m)	Frequency
$0 < m \le 20$	4
$20 < m \le 30$	16
$30 < m \le 40$	14
$40 < m \le 60$	30
$60 < m \le 80$	20
$80 < m \le 120$	14

a On the grid, draw the histogram to illustrate the data in the table.

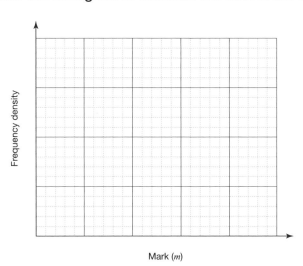

Frequency density

Mark (*m*)

[3 marks]

b Find an estimate for the median mark.

....

[2 marks]

[Total: 5 marks]

23   *VW* and *XY* are a pair of parallel lines.

ABC is an isosceles triangle with *AC* = *BC*.

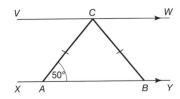

Work out the size of angle *BCW*. You must explain your reasoning.

.... °

[Total: 2 marks]

24   **a** Solve 6(3 − *x*) = 2*x* − 2.

....

[2 marks]

b *n* is an integer with −3 ≤ 2*n* + 1 < 7.

Find all the possible values of *n*.

....

[2 marks]

[Total: 4 marks]

Practice paper (calculator)

Higher tier

Time: 1 hour 30 minutes

The total mark for this paper is 80.
The marks for **each** question are shown in brackets.

1   A stack of 500 sheets of paper is 4.5 cm high.

           Find the thickness of a single sheet of this paper in

           a   centimetres

           .... cm

           [1 mark]

           b   metres, giving your answer in standard form.

           .... m

           [1 mark]

           [Total: 2 marks]

2   Lois is rearranging the formula $T = 2\pi \sqrt{\frac{l}{g}}$ to make l the subject.

           Here is her working.

           $$T = 2\pi\sqrt{\frac{l}{g}}$$

           $$T^2 = 2\pi\frac{l}{g}$$

           $$gT^2 = 2\pi l$$

           $$l = \frac{gT^2}{2\pi}$$

           a   Lois's answer is wrong. Explain why.

           ....

           ....

           [1 mark]

           b   Work out the correct answer.

           ....

           [2 marks]

           [Total: 3 marks]

3   The waiting time in seconds for a call to be answered in a call centre is shown in the table.

Time for call to be answered (t seconds)	Frequency	Cumulative frequency
$0 < t \le 10$	12	
$10 < t \le 20$	18	
$20 < t \le 30$	25	
$30 < t \le 40$	10	
$40 < t \le 50$	2	
$50 < t \le 60$	1	

a Complete the cumulative frequency column.

[2 marks]

b On the grid, draw the cumulative frequency graph.

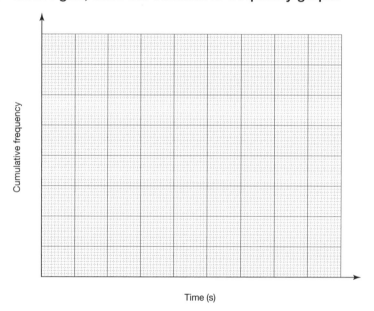

Cumulative frequency

Time (s)

[2 marks]

c Find the median time for a call to be answered.

.... seconds

[1 mark]

d 75% of all calls should be answered within 30 seconds.

Is this target being met? Explain your reasoning.

....

....

....

[2 marks]

[Total: 7 marks]

4   The number of boys in a class is x. The number of girls in the same class is y.

Write the following inequalities in terms of x, y or both.

a   The number of girls is always less than the number of boys in the class.

....

[1 mark]

b   The total number of pupils in the class is always less than 30.

....

[1 mark]

c   The number of boys in the class is never more than double the number of girls in the class.

....

[1 mark]

[Total: 3 marks]

5    $$\frac{1}{R} = \frac{1}{a} + \frac{1}{b}$$

Find R when $a = 6$ and $b = 8$.

Give your answer as a mixed number.

....

[Total: 3 marks]

6   In a class, 10 boys obtained a mean mark of 56% in a maths test.

In the same class, 15 girls obtained a mean mark of 58% in the same maths test.

Claire says that the mean mark for the whole class can be worked out as follows

$$\frac{56 + 58}{2} = 57\%.$$

a   Claire is wrong. Explain why.

....

[1 mark]

b   Work out the correct answer.

....

[2 marks]

[Total: 3 marks]   319

7   The diagram below shows a pipe made of concrete.

The internal radius of the pipe is r m and the external radius is R m. The length of the pipe is h m.

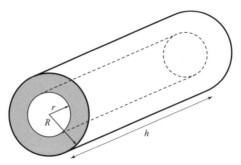

a Show that the shaded area is given by $\pi(R + r)(R - r)$.

[2 marks]

b The internal diameter of the pipe is 1.5 m and the external diameter is 1.6 m.

Find the shaded area in terms of π.

.... m^2

[2 marks]

c The pipe is 4 m long.

The density of concrete is 2400 kg/m³.

Calculate the mass of the pipe.

Give your answer to the nearest whole number.

.... kg

[3 marks]

[Total: 7 marks]

8   A ship sails due north from a port for 40 km. It then turns and travels on a bearing of 040° for 50 km.

   a Calculate the direct distance of the ship from the port.

      Give your answer correct to 1 decimal place.

                                                        .... km

                                                        [2 marks]

   b Calculate the bearing the ship must take for it to sail directly back to the port.

                                                        .... °

                                                        [3 marks]

                                                        [Total: 5 marks]

9   Simplify $(2\sqrt{x} + \sqrt{9y})(\sqrt{x} - 3\sqrt{y})$.

                                                        ....

                                                        [Total: 2 marks]

10   The diagram shows a right-angled triangle.

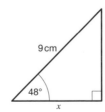

   a Work out the length of the side marked x. Give your answer correct to 1 decimal place.

                                                        ....

                                                        [2 marks]

b Here is a different triangle.

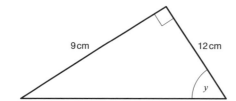

9 cm

12 cm

y

Work out the size of the angle marked y. Give your answer correct to
1 decimal place.

[2 marks]

[Total: 4 marks]

11        $\xi = \{1, 2, 3, 4, 5, 6, 7, 8, 9, 10, 11, 12\}$

F = factors of 24

E = even numbers

a Complete the Venn diagram.

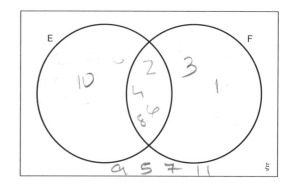

[2 marks]

b One of the numbers is chosen at random.

Find the probability that the number is in

i E ∩ F

[1 mark]

ii (E ∪ F)′

[1 mark]

[Total: 4 marks]

12   Solve $2^x = \frac{1}{8}$.

$x =$ ....

[Total: 2 marks]

13

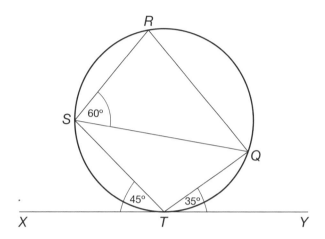

Points Q, R, S and T are on the circumference of a circle.

XY is a tangent to the circle at point T.

Angle $QTY = 35°$, angle $STX = 45°$ and angle $RSQ = 60°$

a   i   Work out the size of angle STQ.

..............................100.............................. °

[1 mark]

    ii   Give a reason for your answer.

..........angles on a straight line add up to 180..........

[1 mark]

b   Work out the size of angle TQR.

    Give a reason for each stage of your working.

.... °

[4 marks]

[Total: 6 marks]

14   During a spell of hot weather, the volume of water in a garden pond decreases due to evaporation.

The volume of water in the pond is initially $30\,m^3$. It decreases at a rate of 2% per week.

a Find the volume of water in the pond after 4 weeks. Give your answer correct to 3 significant figures.

.... m^3

[2 marks]

b If the rate of evaporation remains the same, after how many weeks will the volume of water first fall below half the initial volume?

.... weeks

[3 marks]

[Total: 5 marks]

15   36 staff at a school are planning a night out.

$\frac{4}{9}$ of the staff are male.

Out of the male staff, 25% would like to go to the races and the rest would like to go for a meal.

Of the female staff, $\frac{1}{4}$ would like to go for a meal and the rest would like to go to a concert.

a Use this information to complete the frequency tree.

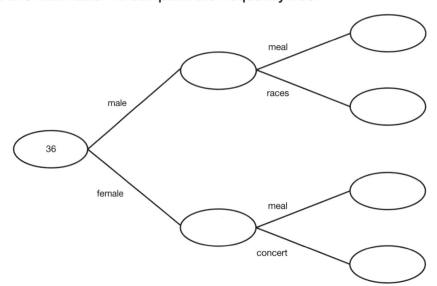

[3 marks]

A member of staff is chosen at random.

b Work out the probability that this member of staff would prefer to go for a meal.

....

[2 marks]

[Total: 5 marks]

16   *VW* and *XY* are parallel lines.

Angle *BAC* = 40°.

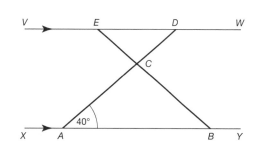

a   i   Work out the size of angle *CDE*.

....°

[1 mark]

ii   Give a reason for your answer.

....

[1 mark]

b Prove that triangles *ABC* and *DEC* are similar.

[2 marks]

c *BC* = 6 cm, *CE* = 4 cm and *AC* = 5 cm.

Work out length *CD*. Give your answer as a mixed number.

.... cm

[2 marks]

[Total: 6 marks]

325

17   Shade the region above the x-axis that is represented by these inequalities.

$x + 3y \leq 24$

$3x + y < 21$

$x \geq 2$

$y \leq 7$

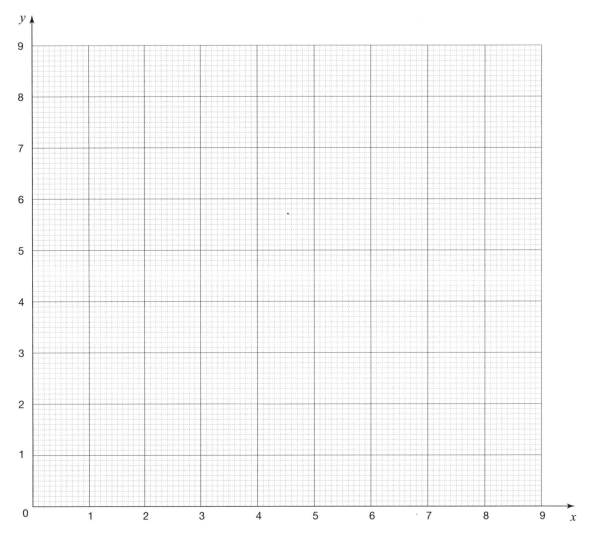

[Total: 5 marks]

18   Duckweed is a plant that rapidly covers the surface of a pond.

In the first week, the area of a pond covered by duckweed increased by 15%.

In the second week, the covered area increased by x%.

In the third week, the covered area also increased by x%.

At the start, the area covered by duckweed was $2\,m^2$.

After 3 weeks, the area covered was $8\,m^2$.

Find the value of x. Give your answer to 2 significant figures.

[Total: 4 marks]

19

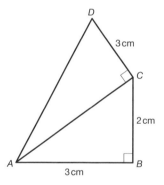

ABC and *ACD* are right-angled triangles.

Show that the length of *AD* is $\sqrt{22}$ cm.

[Total: 4 marks]

Answers

For full worked solutions, visit:
www.scholastic.co.uk/gcse

Number

Integers, decimals and symbols

1   $\frac{1}{0.01}$ 0.1 $(0.1)^2$ $\frac{1}{1000}$ $(-1)^3$
2   **a** 35   **b** 0.01285   **c** −270   **d** 40
3   **a** 4644   **b** 4644   **c** 86   **d** 540
4   **a** $12.56 \times 3.45 = 0.1256 \times 345$
    b $(-8)^2 > -64$ **c** $6 - 12 = 8 - 14$
    d $(-7) \times (0) < (-7) \times (-3)$

Addition, subtraction, multiplication and division

1   **a** 76.765   **b** 201.646   **c** 91.33   **d** 10.564
2   **a** 1176   **c** 44.62   **e** 27
    b 2166   **d** 0.6572   **f** 63
3   **a** 1156   **b** 7.5   **c** 5.76

Using fractions

1   $\frac{2}{5} = \frac{16}{40} = \frac{30}{75} = \frac{50}{125}$
2   **a** $5\frac{1}{3}$ **b** $9\frac{7}{13}$
3   **a** $7\frac{1}{12}$ **b** $7\frac{1}{2}$ **c** $2\frac{9}{20}$
4   $\frac{5}{56}$ 5 $\frac{1}{2}$ $\frac{7}{12}$ $\frac{2}{3}$ $\frac{3}{4}$ $\frac{7}{8}$

Different types of number

1   **a** 7   **b** 49   **c** 2   **d** 6   **e** 6
2   **a** $3^2 \times 7 \times 11$ **b** 63   **c** 10395
3   441   4   5 minutes

Listing strategies

1   210 seconds   3   1100 students
2   5 friends   4   15 pairs

The order of operations in calculations

1   **a** Ravi has worked out the expression from left to right,
        instead of using BIDMAS. He should have performed the
        division and multiplication before the addition.
    b Correct answer: 40
2   **a** 122   **b** −3   **c** 40
3   **a** 6   **b** 14   **c** 8

Indices

1   **a** 10^6 **b** 10^8 **c** 10^6 **d** 10^3
2   **a** 1   **b** $\frac{1}{9}$ **c** 2   **d** 7
3   **a** $\frac{3}{2}$ **b** 16   **c** $\frac{1}{6}$ **d** 64
4   $x = 1.5$

Surds

1   **a** 5   **b** 30   **c** 18
2   $\frac{5\sqrt{3}}{4}$
3   $(2 + \sqrt{3})(2 - \sqrt{3}) = 4 - 2\sqrt{3} + 2\sqrt{3} - 3 = 1$
4   $a = 30$
5   $-\sqrt{5} - 7$
6   $\dfrac{1}{\sqrt{2}} + \dfrac{1}{4} = \dfrac{1 \times \sqrt{2}}{\sqrt{2} \times \sqrt{2}} + \dfrac{1}{4}$
    $= \dfrac{\sqrt{2}}{2} + \dfrac{1}{4}$
    $= \dfrac{2\sqrt{2}}{4} + \dfrac{1}{4}$
    $= \dfrac{1 + 2\sqrt{2}}{4}$

7   $\dfrac{2}{1 - \frac{1}{\sqrt{2}}} = \dfrac{2}{\frac{\sqrt{2}}{\sqrt{2}} - \frac{1}{\sqrt{2}}}$
    $= \dfrac{2}{\frac{\sqrt{2} - 1}{\sqrt{2}}}$
    $= \dfrac{2\sqrt{2}}{\sqrt{2} - 1}$
    $= \dfrac{2\sqrt{2}}{\sqrt{2} - 1} \times \dfrac{\sqrt{2} + 1}{\sqrt{2} + 1}$
    $= \dfrac{4 + 2\sqrt{2}}{2 - 1}$
    $= 4 + 2\sqrt{2}$

8   $\dfrac{3}{\sqrt{3}} + \sqrt{75} + (\sqrt{2} \times \sqrt{6}) = \dfrac{3\sqrt{3}}{3} + \sqrt{3 \times 25} + \sqrt{12}$
    $= \sqrt{3} + 5\sqrt{3} + \sqrt{3 \times 4}$
    $= \sqrt{3} + 5\sqrt{3} + 2\sqrt{3}$
    $= 8\sqrt{3}$

Standard form

1   **a** 2.55×10^{-3} **b** 1.006×10^{10} **c** 8.9×10^{-8}
2   **a** 6×10^{14} **c** 2×10^2 **e** 9×10^{-3}
    b 1.1×10^6 **d** 1×10^{-2}
3   2680   4 $a = 3.3$

Converting between fractions and decimals

1   **a** 0.55   **b** 0.375
2   **a** terminating   **b** recurring   **c** recurring
3   Let $x = 0.\dot{4}0\dot{2} = 0.402402402\ldots$
        $1000x = 402.402402\ldots$
        $1000x - x = 402.402402\ldots - 0.402402402\ldots$
        $999x = 402$
        $x = \dfrac{402}{999} = \dfrac{134}{333}$
        Hence $0.\dot{4}0\dot{2} = \dfrac{134}{333}$
4   $\frac{323}{495}$

Converting between fractions and percentages

1   **a** $\frac{7}{20}$ **b** $\frac{7}{100}$ **c** $\frac{19}{25}$ **d** $\frac{1}{8}$
2   **a** 20%   **b** 68%   **c** 250%   **d** 17.5%
3   53.33% (to 2 d.p.)
4   $\dfrac{66}{90} = \dfrac{66}{90} = 73.3\%$ (to 1 d.p.)
    Jake did better in chemistry.

Fractions and percentages as operators

1   £34.79   4   **a** £14 400   5 $\frac{14}{33}$
2   48   **b** £320
3   7040

Standard measurement units

1   175 000 cm   2   17
3   1286 (to nearest whole number)
4   **a** 1.99×10^{-23} g (to 3 s.f.)   **b** 1.99×10^{-26} kg (to 3 s.f.)
5   7.20×10^{-26} g (to 3 s.f.)

Rounding numbers

1   **a** 35   **c** 0   **e** 2
    b 101   **d** 0
2   **a** 34.88   **b** 34.877
3   **a** 12 800   **b** 0.011   **c** 7×10^{-5}
4   **a** −0.00993 **b** 34.4   **c** 12 300

Estimation

1   200   3   0.16   5   10.6
2   **a** 236.2298627   4   5   6   4
    b 240
7   **a** 5×10^{-28} kg
    b This will be an underestimate, as the mass of one electron
        has been rounded down.

Upper and lower bounds

1   $2.335 \leq l < 2.345$ kg
2   **a** **i** 2.472   **ii** 2.451   **b** 2.5 (to 2 s.f.)
3   34

Algebra

Simple algebraic techniques

1 **a** formula   **c** expression **e** formula
   b identity   **d** identity
2 $x + 6x^2$
3 $y^3 - y = (1)^3 - 1 = 0$ so $y = 1$ is correct.
   $y^3 - y = (-1)^3 - (-1) = -1 + 1 = 0$ so $y = -1$ is correct.
4 **a** $10x$ **b** $4x^2 - 6x$ **c** $18x^2$
5 **a** 2   **b** $\frac{7}{8}$ **c** $-\frac{3}{2}$

Removing brackets

1 **a** $24x - 56$ **b** $-6x + 12$
2 **a** $3x + 9$ **c** $10a^2b - 5ab^2$
   b $8xy + 6x - 2y$ **d** $2x^3y^3 + 3x^2y^4$
3 **a** $m^2 + 5m - 24$ **c** $9x^2 - 6x + 1$
   b $8x^2 + 26x - 7$ **d** $6x^2 + xy - y^2$
4 **a** $x^2 + 7x + 10$ **c** $x^2 - 6x - 7$
   b $x^2 - 16$ **d** $15x^2 + 14x + 3$
5 **a** $x^3 + 6x^2 + 5x - 12$ **b** $18x^3 - 63x^2 + 37x + 20$

Factorising

1 **a** $5x(5x - y)$ **b** $2\pi(2r^2 + 3x)$ **c** $6ab^2(a^2 + 2)$
2 **a** $(3x + 1)(3x - 1)$ **b** $4(2x + 1)(2x - 1)$
3 **a** $(a + 4)(a + 8)$ **b** $(p - 6)(p - 4)$
4 **a** $a(a + 12)$ **b** $(b + 3)(b - 3)$ **c** $(x - 5)(x - 6)$
5 **a** $(3x + 8)(x + 4)$ **b** $(3x + 13)(x - 1)$ **c** $(2x - 5)(x + 2)$
6 $\frac{2}{(x - 3)}$ 7 $\frac{2x - 1}{4x + 1}$

Changing the subject of a formula

1 $T = \frac{PV}{nR}$ 3 $a = \frac{v - u}{t}$ 5 $v = \sqrt{\frac{2E}{m}}$
2 $y = \frac{1 - 4x}{2}$ 4 $x = 5(y + m)$
6 **a** $r = \sqrt{\frac{3V}{\pi h}}$ **b** 3.45 cm (to 2 d.p.)
7 **a** $x = \frac{y + 9}{3}$ **b** 4
8 $x = \frac{3y - 2}{a + 1}$
9 **a** $c = \frac{b}{a}$ **b** upper bound for c = 1.18 (to 3 s.f.)
                              lower bound for c = 1.11 (to 3 s.f.)

Solving linear equations

1 **a** $x = 7$ **d** $x = 32$ **g** $x = -2$
   b $x = 5$ **e** $x = 25$ **h** $x = 84$
   c $x = 4$ **f** $x = -9$
2 $x = \frac{2}{3}$
3 **a** $x = \frac{1}{2}$ **b** $x = -\frac{8}{5}$

Solving quadratic equations using factorisation

1 **a** $(x - 3)(x - 4)$ **b** $x = 3$ or $x = 4$
2 **a** $(2x - 1)(x + 3)$ **b** $x = \frac{1}{2}$ or $x = -3$
3 $x = -2$ or $x = 6$
4 **a** $x(x - 8) - 7 = x(5 - x)$ **b** $x = -\frac{1}{2}$ or $x = 7$
   $x^2 - 8x - 7 = 5x - x^2$
   $2x^2 - 13x - 7 = 0$
5 $x = 2$ cm

Solving quadratic equations using the formula

1 **a** $\frac{3}{x + 7} = \frac{2 - x}{x + 1}$ **b** $x = 1.20$ or -9.20 (to 2 d.p.)
   $3(x + 1) = (2 - x)(x + 7)$
   $3x + 3 = 2x + 14 - x^2 - 7x$
   $3x + 3 = -x^2 - 5x + 14$
   $x^2 + 8x - 11 = 0$
2 $x = 2.78$ cm (to 2 d.p.)   3 $x = 3.30$ or -0.30 (to 2 d.p.)

Solving simultaneous equations

1 $x = 2$ and $y = 3$

2 **a** 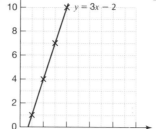   **b** $x = 3, y = 7$

3 $x = 0, y = -3$ or $x = 3, y = 0$

Solving inequalities

1 **a** $x \geq -9$ **b** $x < -12$
2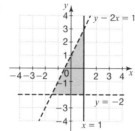
3 **a**

   b (1, 2), (1, 1), (0, 0), (1, 0), (0, -1), (1, -1)
4 $-3 \leq x \leq 1$ 5 $x < -3$ and $x > 5$

Problem solving using algebra

1 42 m^2 2 cost of adult ticket = £7.50
                                          cost of child ticket = £4
3 **a** 16 years   **b** 9 years

Use of functions

1 **a** 19   **b** $x = -1$
2 **a** $(x - 6)^2$ **b** $x^2 - 6$
3 **a** ± 3 **b** $2x + 5$
4 $f^{-1}(x) = \sqrt{\frac{x - 3}{5}}$

Iterative methods

1 Let $f(x) = 2x^3 - 2x + 1$
   $f(-1) = 2(-1)^3 - 2(-1) + 1 = 1$
   $f(-1.5) = 2(-1.5)^3 - 2(-1.5) + 1 = -2.75$
   There is a sign change of $f(x)$, so there is a solution between $x = -1$ and $x = -1.5$.
2 $x_1 = 0.1121111111$
   $x_2 = 0.1125202246$
   $x_3 = 0.1125357073$
3 **a** $x_4 = 1.5213705 \approx 1.521$ (to 3 d.p.)
   b Checking value of $x^3 - x - 2$ for $x = 1.5205, 1.5215$:
       When $x = 1.5205$ $f(1.5205) = -0.0052$
           $x = 1.5215$ $f(1.5215) = 0.0007$
   Since there is a change of sign, the root is 1.521 correct to 3 decimal places.
4 **a i**
   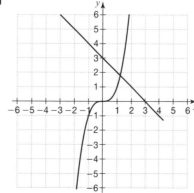

ii There is a root of $x^3 + x - 3 = 0$ where the graphs of $y = x^3$ and $y = 3 - x$ intersect. The graphs intersect once so there is one real root of the equation $x^3 + x - 3 = 0$.

b $x_1 = 1.216440399$
$x_2 = 1.212725591$
$x_3 = 1.213566964$
$x_4 = 1.213376503$
$x_5 = 1.213419623$
$x_6 = 1.213409861 = 1.2134$ (to 4 d.p.)

Equation of a straight line

1 A
2 a $-\frac{4}{3}$ **b** $y = -\frac{1}{2}x + \frac{7}{2}$ **c** $y = 2x + 1$
3 (3.8, 11.4) (to 1 d.p.)

Quadratic graphs

1 a $x = -0.3$ or -3.7 (to 1 d.p.)
b

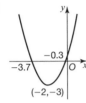

2 $a = 5, b = -2$ and $c = -10$
3 $a = 2, b = 3$ and $c = -15$

Recognising and sketching graphs of functions

1

Equation	Graph
$y = x^2$	B
$y = 2^x$	D
$y = \sin x°$	E
$y = x^3$	C
$y = x^2 - 6x + 8$	A
$y = \cos x°$	F

2 a

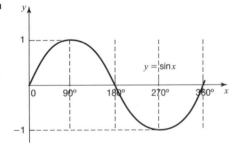

b

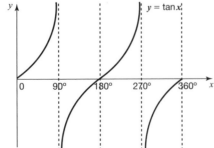

3 $\theta = 70.5°$ or $289.5°$ (to 1 d.p.)

Translations and reflections of functions

1 a

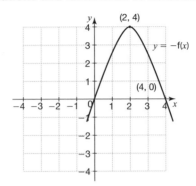

b

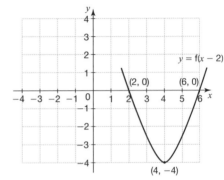

2 a, b, c

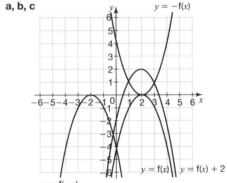

3

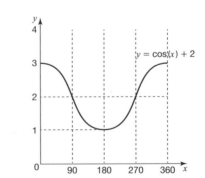

Equation of a circle and tangent to a circle

1 a 5
b 7
c 2
2 radius of the circle $= \sqrt{21} = 4.58$
distance of the point (4, 3) from the centre of the circle (0, 0) $= \sqrt{16 + 9} = \sqrt{25} = 5$
This distance is greater than the radius of the circle, so the point lies outside the circle.
3 a $\sqrt{74}$
b $x^2 + y^2 = 74$ **c** $y = -\frac{5}{7}x + \frac{74}{7}$

Real-life graphs

1 a $1\,\text{m/s}^2$
   b $225\,\text{m}$
2 a The graph is a straight line starting at the origin, so this represents constant acceleration from rest of $\frac{15}{6} = 2.5\,\text{m/s}^2$.
   b The gradient decreases to zero, so the acceleration decreases to zero.
   c $118\,\text{m}$ (to nearest integer); $117\,\text{m}$ is also acceptable
   d It will be a slight underestimate, as the curve is always above the straight lines forming the tops of the trapeziums.

Generating sequences

1 a i $\frac{1}{2}$ ii 243   iii 21
   b 14, 1
2 $-3, -11$
3 a 25, 36
   b 15, 21
   c 8, 13

The nth term

1 a nth term $= 4n - 2$
   b nth term $= 4n - 2 = 2(2n - 1)$
     2 is a factor, so the nth term is divisible by 2 and therefore is even.
   c 236 is not a term in the sequence.
2 a 5
   b -391
   c n^2 is always positive, so the largest value $9 - n^2$ can take is 8 when $n = 1$. All values of n above 1 will make $9 - n^2$ smaller than 8. So 10 cannot be a term.
3 nth term $= n^2 - 3n + 3$

Arguments and proofs

1 The only prime number that is not odd is 2, which is the only even prime number.
Hence, statement is false because 2 is a prime number that is not odd.
2 a true: $n = 1$ is the smallest positive integer and this would give the smallest value of $2n + 1$ which is 3.
   b true: 3 is a factor of $3(n + 1)$ so $3(n + 1)$ must be a multiple of 3.
   c false: $2n$ is always even and subtracting 3 will give an odd number.
3 Let first number $= x$ so next number $= x + 1$
Sum of consecutive integers $= x + x + 1 = 2x + 1$
Regardless of whether x is odd or even, $2x$ will always be even as it is divisible by 2.
Hence $2x + 1$ will always be odd.
4 $(2x - 1)^2 - (x - 2)^2$
$= 4x^2 - 4x + 1 - (x^2 - 4x + 4)$
$= 4x^2 - 4x + 1 - x^2 + 4x - 4$
$= 3x^2 - 3$
$= 3(x^2 - 1)$
The 3 outside the brackets shows that the result is a multiple of 3 for all integer values of x.
5 Let two consecutive odd numbers be $2n - 1$ and $2n + 1$.
$(2n + 1)^2 - (2n - 1)^2$
$= (4n^2 + 4n + 1) - (4n^2 - 4n + 1)$
$= 8n$
Since 8 is a factor of $8n$, the difference between the squares of two consecutive odd numbers is always a multiple of 8.
(If you used $2n + 1$ and $2n + 3$ for the two consecutive odd numbers, difference of squares $= 8n + 8 = 8(n + 1)$.)

Ratio, proportion and rates of change

Introduction to ratios

1 30   3 210 acres   5 144
2 £7500, £8500, £9000   4 $x = \frac{5}{7}$

Scale diagrams and maps

1 5 km
2 a 0.92 km   b 0.12 km
3 1 : 800   4 1 : 200000

Percentage problems

1 10%   3 £18000   5 £896
2 83.3%   4 £16250

Direct and inverse proportion

1 a $P = kT$ b 74074 Pascals (to nearest whole number)
2 £853 (to nearest whole number)
3 a $c = \frac{36}{h}$ b 2.4
4 a €402.50   b £72.07 (to nearest penny)   c £2.50

Graphs of direct and inverse proportion and rates of change

1 B   2 B
3 a
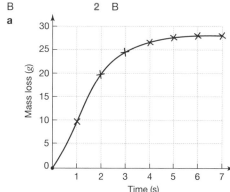
   b i 9.8 g/minute   ii 0.16 g/second (to 2 d.p.)

Growth and decay

1 a 178652   b 5 years
2 £12594
3 0.1 (to 1 s.f.)

Ratios of lengths, areas and volumes

1 a 3.375 or $\frac{27}{8}$ c $133\,\text{cm}^3$ (to nearest whole number)
   b $22.5\,\text{cm}^2$
2 $h = 15\,\text{cm}$ (to nearest cm)
3 a i 9 cm ii 4.5 cm   b 4 : 1

Gradient of a curve and rate of change

1 a $\frac{2}{3}\,\text{m/s}^2$ c $0.37\,\text{m/s}^2$
   b $0.26\,\text{m/s}^2$ d 34 s

Converting units of areas and volumes, and compound units

1 $500\,\text{N/m}^2$
2 $25000\,\text{N/m}^2$
3 107 g (to nearest g)
4 He has worked out the area in m^2 by dividing the area in cm^2 by 100, which is incorrect.
There are $100 \times 100 = 10000\,\text{cm}^2$ in $1\,\text{m}^2$, so the area should have been divided by 10000.
Correct answer:
area in $\text{m}^2 = \frac{9018}{10000}$
$= 0.9018$
$= 0.90\,\text{m}^2$ (to 2 d.p.)
5 72 km/h

Answers

Geometry and measures

2D shapes

1 a true   **c** true   **e** true
b false   **d** true   **f** false (this would be true only for a regular pentagon)

2 a rhombus   **c** equilateral triangle
b parallelogram   **d** kite

Constructions and loci

1

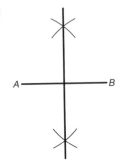

2

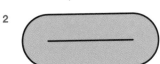

3 a, b

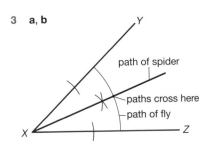

Properties of angles

1 a angle ACB = angle BAC = 30° (base angles of isosceles triangle ABC, since AB and BC are equal sides of a rhombus)

b angle AOB = 90° (diagonals of a rhombus intersect at right angles)

c angle ABO = 180 − (90 + 30) = 60° (angle sum of a triangle)

angle BDC = angle ABO = 60° (alternate angles between parallel lines AB and DC)

2 angle $BAC = \frac{(180 - 36)}{2} = 72°$ (angle sum of a triangle and base angles of an isosceles triangle)

angle BDC = 180 − 90 = 90° (angle sum on a straight line)

angle ABD = 180 − (90 + 72) = 18° (angle sum of a triangle)

3 a $x = 30°$

b If lines AB and CD are parallel, the angles $4x$ and $3x + 30$ would be corresponding angles, and so equal.

$4x = 4 × 30 = 120°$

$3x + 30 = 3 × 30 + 30 = 120°$

These two angles are equal so lines AB and CD are parallel.

4 $x = 90 + 72 = 162°$

Congruent triangles

1 BD common to triangles ABD and CDB
angle ADB = angle CBD (alternate angles)
angle ABD = angle CDB (alternate angles)
Therefore triangles ABD and CDB are congruent (ASA).
Hence angle BAD = angle BCD

2 Draw the triangle and the perpendicular from A to BC.
$AX = AX$ (common)
$AB = AC$ (triangle ABC is isosceles)
angle AXB = AXC = 90° (given)
Therefore triangles ABX and ACX are congruent (RHS).
Hence $BX = XC$, so X bisects BC.

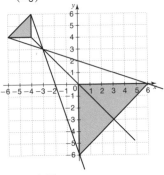

3 OQC = 90° (corresponding angles), so $PB = OQ$ (perpendicular distance between 2 parallel lines)
$AP = PB$ (given), so $AP = OQ$
$PO = QC$ (Q is the midpoint of BC)
angle ABC = angle APO = angle OQC
= 90° (OQ is parallel to AB and OP parallel to BC)
Therefore triangles AOP and OCQ are congruent (SAS).

Transformations

1 translation of $\begin{pmatrix} -7 \\ -5 \end{pmatrix}$

2

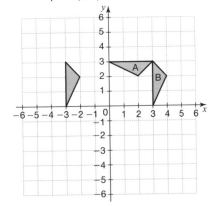

3 a translation of $\begin{pmatrix} -7 \\ 0 \end{pmatrix}$ **b** reflection in the line $y = 3$
c rotation of 90° clockwise about (0, 1)

Invariance and combined transformations

1 a 1
b i invariant point (3, 3)

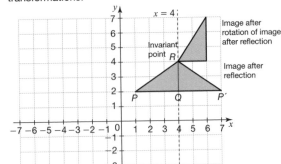

ii rotation 90° anticlockwise about the point (3, 3)

2 a The shaded triangle is the image after the two transformations.

b invariant point is R (4, 4)

3D shapes

1 a G   **c** A, H   **e** C
b B, D   **d** B   **f** A, H

Parts of a circle

1 **a** radius   **c** chord
   b diameter   **d** arc
2 **a** minor sector   **c** major sector
   b major segment **d** minor segment

Circle theorems

1 angle $OTB = 90°$ (angle between tangent and radius)
  angle $BOT = 180 - (90 + 28) = 62°$
  (angle sum in a triangle)
  angle $AOT = 180 - 62 = 118°$
  (angle sum on a straight line)
  $AO = OT$ (radii), so triangle AOT is isosceles
  angle $OAT = \frac{(180 - 118)}{2} = 31°$ (angle sum in a triangle)
2 **a** angle $ACB = 30°$ (angle at centre twice angle at circumference)
  b angle BAC = angle $CBX = 70°$ (alternate segment theorem)
  c $OA = OB$ (radii), so triangle AOB is isosceles
     angle $AOB = 60°$, so triangle AOB is equilateral
     angle $OAB = 60°$ (angle of equilateral triangle)
     angle $CAO = 70 - 60 = 10°$

Projections

1

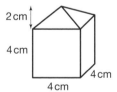

2 **a**

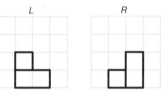

  b

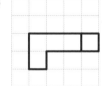

3

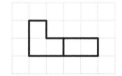

Bearings

1   230°
2

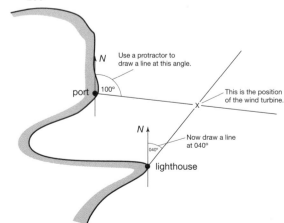

Pythagoras' theorem

1 76 m (to nearest m)
2 9.8 cm (to 1 d.p.)
3 **a** 9.1 cm (to 2 d.p.)
  b 48.76 cm² (to 2 d.p.)

Area of 2D shapes

1 **a** **i** 5.70 cm (to 2 d.p.)
      ii 19.80 cm² (to 2 d.p.)
  b 50.65 cm²
2 **a** 9 cm²   **b** 6 cm²
3 **a** 27π cm²   **b** $18\pi + 6$ cm

Volume and surface area of 3D shapes

1 **a** 3.975 m² **b** 6.36 m³ (to 2 d.p.)
2 5 glasses
3 **a** 7.4 cm (to 1 d.p.)
  b 3.8 cm (to 1 d.p.)
4 0.64 cm

Trigonometric ratios

1 **a** 6.0 cm (to 1 d.p.)
  b 36.9° (to 1 d.p.)
2 44.4° (to 1 d.p.)
3 9.4 cm (to 1 d.p.)
4 21.8° (to 1 d.p.)

Exact values of sin, cos and tan

1

  $\tan 45° = \frac{1}{1} = 1$ $\cos 60° = \frac{1}{2}$

  Hence, $\tan 45° + \cos 60° = 1 + \frac{1}{2} = \frac{3}{2}$

2 **a** **i** $\sin 45° = \frac{1}{\sqrt{2}}$ **ii** $\cos 45° = \frac{1}{\sqrt{2}}$

  b $\frac{\sin 45°}{\cos 45°} = \frac{\frac{1}{\sqrt{2}}}{\frac{1}{\sqrt{2}}} = 1$

    $\tan 45° = 1$

    Hence $\frac{\sin 45°}{\cos 45°} = \tan 45°$

3 **a** $\sqrt{5}$
  b **i** $\sin x = \frac{1}{\sqrt{5}}$ **ii** $\cos x = \frac{2}{\sqrt{5}}$
  c $(\sin x)^2 = \frac{1}{\sqrt{5}} \times \frac{1}{\sqrt{5}} = \frac{1}{5}$

    $(\cos x)^2 = \frac{2}{\sqrt{5}} \times \frac{2}{\sqrt{5}} = \frac{4}{5}$

    $(\sin x)^2 + (\cos x)^2 = \frac{1}{5} + \frac{4}{5}$

                     $= 1$
4 $\tan 30° + \tan 60° + \cos 30°$

  $= \frac{1}{\sqrt{3}} + \sqrt{3} + \frac{\sqrt{3}}{2}$

  $= \frac{\sqrt{3}}{\sqrt{3}\sqrt{3}} + \sqrt{3} + \frac{\sqrt{3}}{2}$

  $= \frac{\sqrt{3}}{3} + \sqrt{3} + \frac{\sqrt{3}}{2}$

  $= \frac{2\sqrt{3} + 6\sqrt{3} + 3\sqrt{3}}{6}$

  $= \frac{11\sqrt{3}}{6}$

Sectors of circles

1 34° (to nearest degree)
2 364.4 cm²
3 **a** 1.67 cm (to 3 s.f.)
  b 1 : 1.04

4 a length of arc $AB = \frac{\theta}{360} \times 2\pi r$

$5.4 = \frac{\theta}{360} \times 2\pi \times 6$

$\theta = 51.5662$

Area of sector $AOB = \frac{\theta}{360} \times 2\pi r^2$

$= \frac{51.5662}{360} \times \pi \times 6^2$

$= 16.2\ cm^2$

Note that both a and b are equal to the radius r of the circle.

b area of triangle AOB

$= \frac{1}{2}ab\sin c$

$= \frac{1}{2} \times 6 \times 6 \sin 51.5662$

$= 14.09988\ cm^2$

area of shaded segment

$=$ area of sector $-$ area of triangle

$= 16.2 - 14.1$

$= 2.1\ cm^2$ (correct to 1 decimal place)

Sine and cosine rules

1 a $225\ cm^2$

b $\frac{4}{5}$

c $18.0\ cm$ (to 3 s.f.)

2 a $18.6\ cm$ (to 3 s.f.)

b $92.4\ cm^2$ (to 3 s.f.)

3 $\frac{2 + 6\sqrt{2}}{17}$

Vectors

1 a $\begin{pmatrix} 2 \\ 1 \end{pmatrix}$ b $\begin{pmatrix} -22 \\ 9 \end{pmatrix}$

2 a $\mathbf{b} - \mathbf{a}$ b $\frac{3}{5}(\mathbf{b} - \mathbf{a})$

c $\overrightarrow{OQ} = \frac{2}{5}\overrightarrow{OA} = \frac{2}{5}\mathbf{a}$

$\overrightarrow{QP} = \overrightarrow{QA} + \overrightarrow{AP}$

$= \frac{3}{5}\mathbf{a} + \frac{3}{5}(\mathbf{b} - \mathbf{a})$

$= \frac{3}{5}\mathbf{a} + \frac{3}{5}\mathbf{b} - \frac{3}{5}\mathbf{a}$

$= \frac{3}{5}\mathbf{b}$

As $\overrightarrow{QP} = \frac{3}{5}\mathbf{b}$ and $\overrightarrow{OB} = \mathbf{b}$ they both have the same vector part and so are parallel.

3 a $\overrightarrow{BC} = \overrightarrow{BA} + \overrightarrow{AC}$

$= -3\mathbf{b} + \mathbf{a}$

$= \mathbf{a} - 3\mathbf{b}$

b $\overrightarrow{PB} = \frac{1}{3}\overrightarrow{AB} = \mathbf{b}$

$\overrightarrow{PM} = \overrightarrow{PB} + \overrightarrow{BM}$

$= \overrightarrow{PB} + \frac{1}{2}\overrightarrow{BC}$

$= \mathbf{b} + \frac{1}{2}(\mathbf{a} - 3\mathbf{b})$

$= \frac{1}{2}\mathbf{a} - \frac{1}{2}\mathbf{b}$

$= \frac{1}{2}(\mathbf{a} - \mathbf{b})$

$\overrightarrow{MD} = \overrightarrow{MC} + \overrightarrow{CD}$

$= \frac{1}{2}\overrightarrow{BC} + \overrightarrow{CD}$

$= \frac{1}{2}(\mathbf{a} - 3\mathbf{b}) + \mathbf{a}$

$= \frac{3}{2}\mathbf{a} - \frac{3}{2}\mathbf{b}$

$= \frac{3}{2}(\mathbf{a} - \mathbf{b})$

Both $\overrightarrow{PM}$ and $\overrightarrow{MD}$ have the same vector part $(\mathbf{a} - \mathbf{b})$ so they are parallel. Since they both pass through M, they are parts of the same line, so PMD is a straight line.

Probability

The basics of probability

1 a

		Dice 1				
		1	2	3	4	5	6
	1	2	3	4	5	6	7
	2	3	4	5	6	7	8
	3	4	5	6	7	8	9
Dice 2	4	5	6	7	8	9	10
	5	6	7	8	9	10	11
	6	7	8	9	10	11	12

b $\frac{1}{36}$ c $\frac{5}{12}$ d 7

2 a 21 chocolates   b $\frac{3}{7}$

3

		Bethany				
		1	2	3	4	5	6
	1	1, 1	1, 2	1, 3	1, 4	1, 5	1, 6
	2	2, 1	2, 2	2, 3	2, 4	2, 5	2, 6
	3	3, 1	3, 2	3, 3	3, 4	3, 5	3, 6
Amy	4	4, 1	4, 2	4, 3	4, 4	4, 5	4, 6
	5	5, 1	5, 2	5, 3	5, 4	5, 5	5, 6
	6	6, 1	6, 2	6, 3	6, 4	6, 5	6, 6

a $\frac{1}{6}$ b $\frac{5}{12}$

Probability experiments

1 a He is wrong because 100 spins is a very small number of trials. To approach the theoretical probability you would have to spin many more times. Only when the number of spins is extremely large will the frequencies start to become similar.

b $\frac{11}{50}$ c 95

2 a $x = 0.08$ b 0.24   c 16

The AND and OR rules

1 a $\frac{9}{169}$ b $\frac{3}{169}$ c $\frac{1}{52} \times \frac{1}{52} = \frac{1}{2704}$

2 a When an event has no effect on another event, they are said to be independent events. Here the colour of the first marble has no effect on the colour of the second marble.

b $\frac{9}{100}$ c $\frac{3}{10}$

3 a $\frac{9}{140}$ b $\frac{6}{35}$

Tree diagrams

1 a $\frac{2}{15}$ b $\frac{8}{15}$

2 a

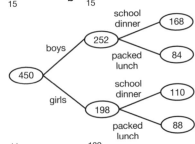

b $\frac{11}{45}$ c $\frac{139}{225}$

3 a 0.179 (to 3 d.p.)   b 0.238 (to 3 d.p.)   c 0.131 (to 3 d.p.)

Venn diagrams and probability

1 a 9, 8   c 1, 3, 4, 10, 12, 15

b 1, 2, 3, 5, 7, 8, 9, 12, 15   d 4, 10

2 a

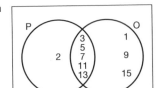

          b $\frac{1}{3}$

3 a $\frac{17}{20}$ b $\frac{18}{73}$

Statistics

Sampling

1 a Ling's, as he has a larger sample so it is more likely to represent the whole population (i.e. students at the school).
  b 22
2 The sample should be taken randomly, with each member of the population having an equal chance of being chosen.
  The sample size should be large enough to represent the population, since the larger the sample size, the more accurate the results.
3 a 173
  b 25

Two-way tables, pie charts and stem-and-leaf diagrams

1 a

	Coronation Street	EastEnders	Emmerdale	Total
Boys	12	31	20	63
Girls	18	12	7	37
Total	30	43	27	100

  b $\frac{27}{100}$
  c $\frac{12}{37}$

2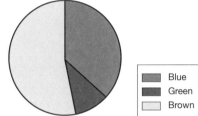

    Blue
    Green
    Brown

3 a

Number of students late

```
0 | 3  4  4  5  8  8  9
1 | 0  0  1  1  2  2  2  2  6  7  8
2 | 0  2  6
```

Key: 1|2 means 12 students late

  b 11 students
  c 23 students

Line graphs for time series data

1 a January to March: $\frac{12 + 10 + 20}{3} = 14$

  February to April: $\frac{10 + 20 + 54}{3} = 28$

  March to May: $\frac{20 + 54 + 87}{3} = 53.6$

  April to June: $\frac{54 + 87 + 130}{3} = 90.3$

  b increasing sales

2 a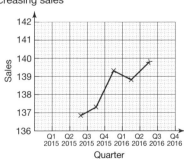

  b The general trend is that the sales are increasing.

Averages and spread

1 a 32
  b 2.7 (to 1 d.p.)
  c 3
2 66.5 (to 1 d.p.)
3 a

Age (t years)	Frequency	Mid-interval value	Frequency × mid-interval value
$0 < t \leq 4$	8	2	16
$4 < t \leq 8$	10	6	60
$8 < t \leq 12$	16	10	160
$12 < t \leq 14$	1	13	13

  b 35
  c 7.1 years (to 2 s.f.)

Histograms

1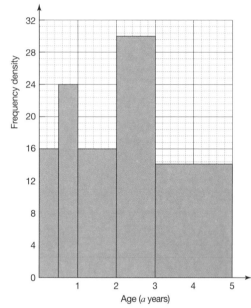

2

Wingspan (w cm)	Frequency
$0 < w \leq 5$	2
$5 < w \leq 10$	6
$10 < w \leq 15$	24
$15 < w \leq 25$	30
$25 < w \leq 40$	9

Cumulative frequency graphs

1 a $0 < n \leq 2$
  b

Number of items of junk mail per day (n)	Frequency	Cumulative frequency
$0 < n \leq 2$	21	21
$2 < n \leq 4$	18	39
$4 < n \leq 6$	10	49
$6 < n \leq 8$	8	57
$8 < n \leq 10$	3	60
$10 < n \leq 12$	1	61

c

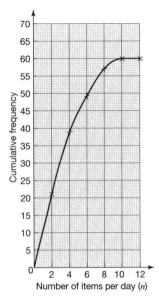

Number of items per day (n)

d median = value at frequency of 30.5 = 3

2 a i 15.8 kg **ii** 6.5 kg

    b 32 penguins

Comparing sets of data

1 a

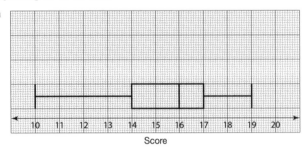

Score

    b Sasha's median score is higher (17 compared to 16). The IQR for Sasha is 2 compared to Chloe's 3. The range for Sasha is 5 compared to Chloe's 9. Both these are measures of spread, which means that Sasha's scores are less spread out (i.e. more consistent).

2 a   i 138 minutes **ii** 100 minutes **iii** 22 minutes

    b On average the men were faster as the median is lower.

       The variation in times was greater for the men as their range was greater, although the spread of the middle half of the data (the interquartile range) was slightly greater for the women.

Scatter graphs

1 a, b

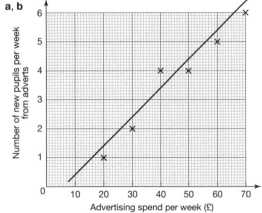

Advertising spend per week (£)

    c positive correlation

2 a, b

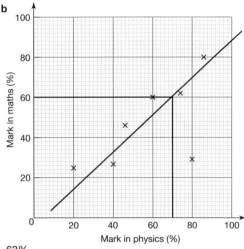

Mark in physics (%)

    c 63%

Practice papers

Non-calculator

1 29.382

2 a $3\frac{1}{3}$

    b $2\frac{11}{40}$

3 a 1.4094

    b 4.86

4 $a^2 + 12a + 27$

5 $4(y + 4)(y - 4)$

6 15:3:1

7 a $8x^6y^3$

    b $6x$

    c $\frac{5}{b}$

8 a $a = 1, b = 5$

    b i (1, 5)

       ii

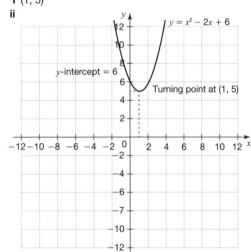

y-intercept = 6

Turning point at (1, 5)

$y = x^2 - 2x + 6$

       iii The curve does not intersect the x-axis (i.e. where $y = 0$), so the equation has no roots.

9 a 10

    b $\frac{16}{25}$

10 $y = \frac{3 - 2c}{2a + 1}$

11 364 cm³ (to nearest integer)

12 $\frac{\sqrt{3} - 2}{\sqrt{3} + 1} = \frac{\sqrt{3} - 2}{\sqrt{3} + 1} \times \frac{(\sqrt{3} - 1)}{(\sqrt{3} - 1)}$

$$= \frac{3 - 3\sqrt{3} + 2}{3 - 1}$$

$$= \frac{5 - 3\sqrt{3}}{2}$$

13 Let $x = 0.13\dot{6} = 0.136363636...$
$$10x = 1.36363636...$$
$$1000x = 136.363636$$
$$1000x - 10x = 136.363636...$$
$$- 1.36363636...$$
$$990x = 135$$
$$x = \frac{135}{990} = \frac{3}{22}$$

14 a $(0, -6)$ **b** $(0, 6)$

15 a

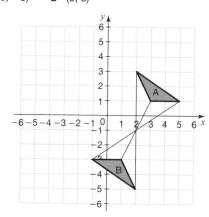

b A rotation of 180° (clockwise or anticlockwise) about the point $(2, -1)$.

16 $x^2 \leq 2x + 15$

$x^2 - 2x + 15 \leq 0$

Factorising $x^2 - 2x - 15 = 0$ gives $(x - 5)(x + 3) = 0 : x = 5$ or -3

As the coefficient of x^2 is positive, the graph of $y = x^2 - 2x - 15$ is ∪-shaped.

$x^2 - 2x - 15 \leq 0$ for the region below the x-axis (i.e. where $y \leq 0$).

$-3 \leq x \leq 5$

17 a

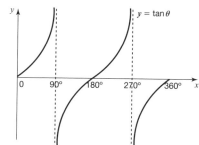

b $\theta = 45°$ or $225°$

18 $3x + 4y - 25 = 0$

19 $x = 5$ and $y = 1$

20 $3n^2 - 2n + 4$

21 a i constant speed of 4 m/s

ii constant acceleration for 3 seconds

b 10 m/s

22 a

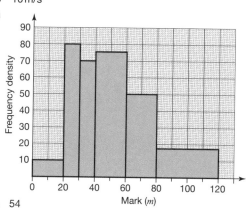

b 54

23 angle $ABC = 50°$ (base angles of an isosceles triangle equal)
angle $BCW = 50°$ (alternate angles)

24 a $x = 2.5$ **b** $n = -2, -1, 0, 1, 2$

Calculator

1 a 0.009 cm

b 9×10^{-5} m

2 a When Lois squared both sides to remove the square root sign, she did not square the 2π.

b $l = \frac{gT^2}{4\pi^2}$

3 a

Time for call to be answered (t seconds)	Frequency	Cumulative frequency
$0 < t \leq 10$	12	12
$10 < t \leq 20$	18	30
$20 < t \leq 30$	25	55
$30 < t \leq 40$	10	65
$40 < t \leq 50$	2	67
$50 < t \leq 60$	1	68

b

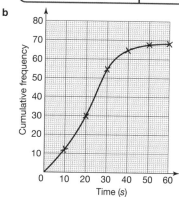

c 21 seconds

d 75% of 68 = 0.75 × 68 = 51

Reading off from 51 on the cumulative frequency axis gives a time of 28 s.

75% of calls are answered within 28 s, so the target is being met.

4 a $y < x$

b $x + y < 30$

c $x \leq 2y$

5 $3\frac{3}{7}$

6 a There are different numbers of boys and girls so this needs to be taken into account when the mean of the whole class is found.

b 57.2%

7 a shaded area = area of large circle − area of small circle
$$= \pi R^2 - \pi r^2$$
$$= \pi(R^2 - r^2)$$
$$= \pi(R + r)(R - r)$$

b 0.0775π m²

c 9349 kg (to nearest whole number)

8 a 84.6 km (to 1 d.p.)

b 202°

9 $2x - 3\sqrt{xy} - 9y$

10 a 6.0 cm (to 1 d.p.)

b 36.9° (to 1 d.p.)

11 a

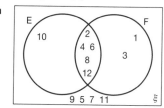

b i $\frac{5}{12}$ ii $\frac{1}{3}$

12 $x = -3$

13 **a** **i** 100°

      ii Angles on a straight line sum to 180°.

    b angle $TSQ = 35°$ (Angle in the alternate segment)

      angle $TSR = 60 + 35 = 95°$

      angle $TQR = 180 - 95 = 85°$ (opposite angles of a cyclic
                                        quadrilateral add up to 180°)

14 **a** 27.7 m³ (to 3 s.f.)   **b** 35 weeks

15 **a**

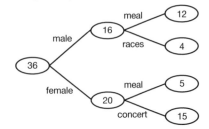

    b $\frac{17}{36}$

16 **a** **i** angle $CDE = 40°$ **ii** alternate angle to angle BAD

    b Angle ABC = angle CED as they are alternate angles.

      Angle DCE = angle BCA as vertically opposite angles
      are equal.

      All the equivent angles in both triangles are the same so
      the triangles are similar.

    c $3\frac{1}{3}$ cm

17

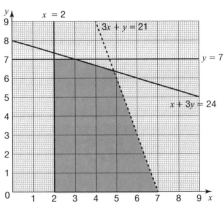

18 87 % (to 2 s.f.)

19 Using Pythagoras' theorem in triangle ABC:

    $AC^2 = 2^2 + 3^2 = 13$

    $AC = \sqrt{13}$ cm

    By Pythagoras' theorem in triangle ACD:

    $AD^2 = \left(\sqrt{13}\right)^2 + 3^2 = 13 + 9 = 22$

    $AD = \sqrt{22}$ cm

Notes

Notes